# AGROECOSYSTEM SUSTAINABILITY
## Developing Practical Strategies

# Advances in Agroecology
Series Editor: Clive A. Edwards

# AGROECOSYSTEM SUSTAINABILITY
## Developing Practical Strategies

By

# Stephen R. Gliessman

**CRC Press**
Taylor & Francis Group
Boca Raton London New York

CRC Press is an imprint of the
Taylor & Francis Group, an **informa** business

CRC Press
Taylor & Francis Group
6000 Broken Sound Parkway NW, Suite 300
Boca Raton, FL 33487-2742

First issued in paperback 2019

ISBN-13: 978-0-8493-0894-9 (hbk)
ISBN-13: 978-0-367-39811-8 (pbk)

| Library of Congress Cataloging-in-Publication Data |
| --- |
| Gliessman, Stephen R.<br>    Agroecosystem sustainability : developing practical strategies / Stephen Gliessman.<br>        p.    cm.<br>    Includes bibliographical references (p.).<br>    ISBN 0-8493-0894-1 (alk. paper)<br>    1. Agricultural ecology. 2. Sustainable agriculture. I. Title.<br><br>S589.7 .G584 2000<br>630'.2'77—dc21                                                    00-056485<br>                                                                        CIP |

Library of Congress Card Number 00-056485

Visit the Taylor & Francis Web site at
http://www.taylorandfrancis.com

and the CRC Press Web site at
http://www.crcpress.com

# Preface

Considerable evidence indicates that modernized, conventional agroecosystems around the world are unsustainable. Dependent on large, fossil-fuel-based, external inputs, they are overusing and degrading the soil, water, and genetic resources upon which agriculture depends. Although the deterioration of agriculture's foundation can be masked by fertilizers, herbicides, pesticides, high-yielding varieties, and water and fossil-fuel resources borrowed from future generations, it cannot be hidden forever, especially given increases in the human population, climate modification, and destruction of natural biodiversity and habitats.

It is against this background of concern that the science of agroecology and the concept of sustainability have arisen and evolved during recent decades. Agroecological research has always held sustainability of food production systems as its ultimate goal; recently agroecological and related research have turned toward making its connection to sustainability stronger and working on more practical strategies for shifting toward sustainability in agriculture.

This volume showcases the leading research in developing practical strategies. This research ranges from specific management practices that can enhance agroecosystem sustainability in a region to more global efforts to develop sets of sustainability indicators that can assess movement toward or away from sustainability.

Although the chapters in this volume represent disparate levels of focus and various disciplinary approaches, each chapter is part of the larger puzzle of achieving sustainability in agriculture, and springs from an agroecological framework. Modern agroecosystems have become unsustainable for a variety of reasons having to do with economics, history, social and political change, and the nature of technological development. Redirecting agriculture in a sustainable direction requires research and change in all these areas, but the basis of sustainability lies in ecological understanding of agroecosystems dynamics as represented by agroecology.

The chapters in this volume are organized into three sections: The first section presents the results of research in specific strategies for increasing the sustainability of farming systems. Particular problems or conditions facing farm managers are identified, and alternatives that employ an agroecological framework are applied. These strategies include adding self-reseeding annual legumes to a conventional crop rotation, manipulating the spatial distribution of natural biodiversity in vineyards to enhance natural pest control, applying agroforestry practices, and managing mulch.

The second section presents a variety of research approaches for assessing the level or degree of sustainability of farming systems. Each chapter in this section focuses on a particular agroecosystem level process or condition — ranging from nematode communities in the soil to nutrient cycling — that can be used to evaluate performance and sustainability as a function of farm design and management.

The third section takes sustainability analysis to its most holistic level through the presentation of research that combines the ecological foundations of sustainability with their social components. These chapters attempt to place agroecology in the social and cultural environment in order to influence people's decisions on how and why to design and manage agroecosystems.

Ultimately, this book emphasizes sustainability as a whole-system, interdisciplinary concept, and that it is the emergent quality of agroecosystems that evolves over time. Sustainability is the integration of a recognizable social system and its ecosystem setting; it results in a dynamic, continually evolving agroecosystem.

Stephen R. Gliessman

# The Editor

With graduate degrees in botany, biology, and plant ecology from the University of California, Santa Barbara, Stephen R. Gliessman has over 25 years of teaching, research, and production experience in the field of agroecology. He has hands-on and academic experience in tropical to temperate agriculture, small farm to large farm systems, traditional to conventional farm management, and organic and synthetic chemical approaches to agroecosystem design and management. He is the founding director of the University of California, Santa Cruz Agroecology Program (one of the first formal agroecology programs in the world), and is the Alfred Heller Professor of Agroecology in the Department of Environmental Studies at UCSC. He dry farms organic wine grapes and olives with his brother in northern Santa Barbara County, California.

# Contributors

**Miguel A. Altieri**
ESMP, Division of Insect Biology
University of California, Berkeley
201 Wellman-3112
Berkeley, CA 94720-3112
*(agroeco3@nature.berkeley.edu)*

**Enio Campligia**
Dipartimento di Produzione Vegetale
Universita degli Studi della Tuscia
Via S. Camillo de Lellis
01100 Viterbo, Italy
*(campligi@unitus.it)*

**Fabio Caporali**
Dipartimento di Produzione Vegetale
Universita degli Studi della Tuscia
Via S. Camillo de Lellis
01100 Viterbo, Italy
*(caporali@unitus.it)*

**Xu Cheng**
Department of Agronomy and
  Agroecology
China Agricultural University
Beijing 100094
P.R. China
*(chengxu@public.east.cn.net)*

**Erle C. Ellis**
Center for Agroecology and Sustainable
  Food Systems
University of California, Santa Cruz
Santa Cruz, CA 95064
*(ece@umbc.edu)*

**Abbas Farshad**
International Institute for Aerospace
  Survey and Earth Sciences
Soil Science Division
7500 AA Enschede, The Netherlands
*(Farshad@itc.nl)*

**Remi Gauthier**
Environment Department
Wye College, University of London
Wye, Ashford
Kent TN25 5AH
England, UK
*(R.Gauthier@wye.ac.uk)*

**Mario Giampietro**
Istituto Nazionale Ricerche su Alimenti
  e Nutrizione
Unit of Technological Assessment
Via Ardeatina 546
00178 Rome, Italy
*(giampietro@inn.ingrm.it)*

**Stephen R. Gliessman**
Alfred Heller Professor of Agroecology
Department of Environmental Studies
University of California, Santa Cruz
Santa Cruz, CA 95064
*(gliess@zzyx.ucsc.edu)*

**Rong Gang Li**
Office of Rural Energy and
  Environmental Protection
Jiangsu Department of Agriculture and
  Forestry
Nanjing, Jiangsu 210009
P. R. China

**Zhengfang Li**
Intercontinental Center for
  Agroecological Industry Research and
  Development
Nanjing
P. R. China
*(icaird@jlonline.com)*

**Rodrigo M. Machado**
Departamento de Biologia General
Instituto de Ciências Biologicas
Universidade Federal de Minas Gerais
Belo Horizonte, Brazil

**V. Ernesto Méndez**
Department of Environmental Studies
University of California, Santa Cruz
Santa Cruz, CA 95064
*(vemendez@cats.ucsc.edu)*

**Joji Muramoto**
Center for Agroecology and Sustainable
  Food Systems
University of California, Santa Cruz
Santa Cruz, CA 95064
*(joji@cats.ucsc.edu)*

**Deborah A. Neher**
Department of Biology
University of Toledo
Toledo, OH 43606
*(dneher@uoft02.utoledo.edu)*

**Clara Nicholls**
ESPM, Division of Insect Biology
University of California, Berkeley
Berkeley, CA 94720
*(agroeco3@nature.berkeley.edu)*

**Gianni Pastore**
Istituto Nazionale Ricercha su Alimenti
  e Nutrizione
Via Ardeatina 546
00178 Rome, Italy
*(pastore@inn.ingrm.it)*

**Martha E. Rosemeyer**
Department of Agronomy
University of Wisconsin
Madison, WI 53706
*(merosemeyer@facstaff.wisc.edu)*

**Graham Woodgate**
Environment Department
Wye College, University of London
Wye, Ashford
Kent TN25 5AH
England, UK
*(G.Woodgate@wye.ac.uk)*

**Lin Zhang Yang**
Department of Ecology
Nanjing Institute of Soil Sciences
Chinese Academy of Sciences
Nanjing, Jiangsu 210008
P.R. China
*(lzyang@mail.issas.ac.cn)*

**Joseph A. Zinck**
International Institute for Aerospace
  Survey and Earth Sciences
P. O. Box 6
7500 AA Enschede, The Netherlands

# Contents

# Section I
# Increasing Sustainability

# CHAPTER 1

# The Ecological Foundations of Agroecosystem Sustainability*

Stephen R. Gliessman

## CONTENTS

## 1.1 INTRODUCTION

What is a sustainable agroecosystem? An easy way to answer this question is to give a definition: A sustainable agroecosystem maintains the resource base upon which it depends, relies on a minimum of artificial inputs from outside the farm system, manages pests and diseases through internal regulating mechanisms, and is able to recover from the disturbances caused by cultivation and harvest (Edwards et al., 1990; Altieri, 1995). Such a broadly applicable definition still begs many other questions: How do we identify an actually existing agroecosystem as sustainable or

* This chapter is adapted from Chapter 20 of *Agroecology: Ecological Processes in Sustainable Agriculture*, by Stephen Gliessman, CRC Press LLC, Boca Raton, FL, 2000.

not? What particular facets of a system make it sustainable or unsustainable? How can we build a sustainable system in a particular bioregion, given realistic economic constraints? Generating the knowledge and expertise for answering these kinds of questions is one of the main tasks facing the science of agroecology today.

Ultimately, sustainability is a test of time; an agroecosystem that has continued to be productive for a long period of time without degrading its resource base — either locally or elsewhere — can be said to be sustainable. What constitutes a long period of time? How is it determined if degradation of resources has occurred? How can a sustainable system be designed when the proof of its sustainability remains always in the future?

Despite these challenges, we need to determine what sustainability entails. In short, the task is to identify parameters of sustainability — specific characteristics of agroecosystems that play key parts in agroecosystem function — and to determine at what level or condition these parameters must be maintained for sustainable function to occur. Through this process, we can identify what we will call indicators of sustainability — agroecosystem-specific conditions necessary for and indicative of sustainability. With such knowledge it will be possible to predict whether a particular agroecosystem can be sustained over the long-term, and to design agro-ecosystems that have the best chance of proving to be sustainable.*

## 1.2 LEARNING FROM EXISTING SUSTAINABLE SYSTEMS

The process of identifying the elements of sustainability begins with two kinds of existing systems: natural ecosystems and traditional agroecosystems. Both have stood the test of time in terms of maintaining productivity over long periods, and each offers a different kind of knowledge foundation. Natural ecosystems provide an important reference point for understanding the ecological basis of sustainability; traditional agroecosystems offer abundant examples of actually sustainable agricultural practices as well as insights into how social systems — cultural, political, and economic — fit into the sustainability equation. Based on the knowledge gained from these systems, agroecological research can devise principles, practices, and designs that can be applied in converting unsustainable conventional agroecosystems into sustainable ones.

### 1.2.1 Natural Ecosystems as Reference Points

Natural ecosystems and conventional agroecosystems are very different. Conventional agroecosystems are generally more productive but far less diverse than natural systems. Unlike natural systems, conventional agroecosystems are far from self-sustaining. Their productivity can be maintained only with large additional inputs of energy and materials from external, human sources; otherwise they quickly degrade to a much less productive level. In every respect, these two types of systems are at opposite ends of a spectrum.

The key to sustainability is to find a compromise between a system that models the structure and function of natural ecosystems and yields a harvest for human use.

---

* For recent reviews of different ways to apply sustainability analysis see Munasinghe and Shearer 1995; Moldan et al., 1997; OCED, 1998.

Such a system is manipulated to a high degree by humans for human ends, and is therefore not self-sustaining, but relies on natural processes for maintenance of its productivity. Its resemblance to natural systems allows the system to sustain human appropriation of its biomass without large subsidies of industrial cultural energy and detrimental effects on the surrounding environment.

Table 1.1 compares these three types of systems using several ecological criteria. As the terms in the table indicate, sustainable agroecosystems model the high diversity, resilience, and autonomy of natural ecosystems. Compared to conventional systems, they have somewhat lower and more variable yields, a reflection of the variation that occurs from year to year in nature. These lower yields, however, are usually more than offset by the advantage gained in reduced dependence on external inputs and an accompanying reduction in adverse environmental impacts.

From this comparison we can derive a general principle: the greater the structural and functional similarity of an agroecosystem to the natural ecosystems in its biogeographic region, the greater the likelihood that the agroecosystem will be sustainable. If this principle holds true, then observable and measurable values for a range of natural ecosystem processes, structures, and rates can provide threshold values or benchmarks that delineate the ecological potential for the design and management of agroecosystems. It is the task of research to determine how close an agroecosystem needs to be to these benchmark values to be sustainable (Gliessman, 1990).

## 1.2.2 Traditional Agroecosystems as Examples of Sustainable Function

Throughout much of the rural world today, traditional agricultural practices and knowledge continue to form the basis for much of the primary food production.

Table 1.1   Properties of Natural Ecosystems, Sustainable Agroecosystems, and Conventional Agroecosystems

|  | Natural Ecosystems | Sustainable Agroecosystems[a] | Conventional Agroecosystems[a] |
|---|---|---|---|
| Production (yield) | Low | low/medium | high |
| Productivity (process) | Medium | medium/high | low/medium |
| Species diversity | High | medium | low |
| Resilience | High | medium | low |
| Output stability | Medium | low/medium | high |
| Flexibility | High | medium | low |
| Human displacement of ecological processes | Low | medium | high |
| Reliance on external human inputs | Low | medium | high |
| Internal nutrient cycling | High | medium/high | low |
| Sustainability | High | high | low |

[a] Properties given for these systems are most applicable to the farm scale and for the short to medium term time frame.

Modified from Odum (1984), Conway (1985), and Altieri (1995).

What distinguishes traditional and indigenous production systems from conventional systems is that the former developed primarily in times or places where inputs other than human labor and local resources were not available, or where alternatives have been found that reduce, eliminate, or replace the energy- and technology-intensive human inputs common to much of present-day conventional agriculture. The knowledge embodied in traditional systems reflects experience gained from past generations, yet continues to develop in the present as the ecological and cultural environment of the people involved go through the continual process of adaptation and change (Wilken, 1988).

Many traditional farming systems can allow for the satisfaction of local needs while also contributing to food demands on the regional or national level. Production takes place in ways that focus more on the long-term sustainability of the system, rather than solely on maximizing yield and profit. Traditional agroecosystems have been in use for a long time, and have gone through many changes and adaptations. The fact that they still are in use is strong evidence for social and ecological stability that modern, mechanized systems could well envy (Klee, 1980).

Studies of traditional agroecosystems can contribute greatly to the development of ecologically sound management practices. Indeed, our understanding of sustainability in ecological terms comes mainly from knowledge generated from such study (Altieri, 1990).

What are the characteristics of traditional agroecosystems that make them sustainable? Despite the diversity of these agroecosystems across the globe, we can begin to answer this question by examining what most traditional systems have in common. Traditional agroecosystems:

- Do not depend on external, purchased inputs
- Make extensive use of locally available and renewable resources
- Emphasize the recycling of nutrients
- Have beneficial or minimal negative impacts on both the on and off  farm environment
- Are adapted to or tolerant of local conditions, rather than dependent on massive alteration or control of the environment
- Are able to take advantage of the full range of microenvironmental variation within the cropping system, farm, and region
- Maximize yield without sacrificing the long-term productive capacity of the entire system and the ability of humans to use its resources optimally
- Maintain spatial and temporal diversity and continuity
- Conserve biological and cultural diversity
- Rely on local crop varieties and often incorporate wild plants and animals
- Use production to meet local needs first
- Are relatively independent of external economic factors
- Are built on the knowledge and culture of local inhabitants

Traditional practices cannot be transplanted directly into regions of the world where agriculture has already been modernized, nor can conventional agriculture be converted to fit the traditional mold exactly. Nevertheless, traditional practices hold important lessons for how modern sustainable agroecosystems should be designed.

A sustainable system need not have all the characteristics outlined above, but it must be designed so that all the functions of these characteristics are retained.

Traditional agroecosystems can provide important lessons about the role that social systems play in sustainability. For an agroecosystem to be sustainable, the cultural and economic systems in which its human participants are embedded must support and encourage sustainable practices and not create pressures that undermine them. The importance of this connection is revealed when formerly sustainable traditional systems undergo changes that make them unsustainable or environmentally destructive. In every case, the underlying cause is some kind of social, cultural, or economic pressure. For example, it is a common occurrence for traditional farmers to shorten fallow periods or increase their herds of grazing animals in response to higher rents or other economic pressures and to have these changes cause soil erosion or reduction in soil fertility.

It is essential that traditional agroecosystems be recognized as examples of sophisticated, applied ecological knowledge. Otherwise, the so called modernization process in agriculture will continue to destroy the time tested knowledge they embody — knowledge that should serve as a starting point for the conversion to the more sustainable agroecosystems of the future.

## 1.3 CONVERTING TO SUSTAINABLE PRACTICES

Farmers have a reputation for being innovators and experimenters, willing to adopt new practices when they perceive some benefit will be gained. Over the past 40 to 50 years, innovation in agriculture has been driven mainly by an emphasis on high yields and farm profit, resulting in remarkable returns but also an array of negative environmental side effects. Despite the continuation of this strong economic pressure on agriculture, however, many farmers are choosing to make the transition to practices that are more environmentally sound and have the potential for contributing to long-term sustainability for agriculture (National Research Council, 1989; Edwards et al., 1990).

Several factors are encouraging farmers to enter into this transition process:

- The rising cost of energy
- The low profit margins of conventional practices
- The development of new practices that are seen as viable options
- Increasing environmental awareness among consumers, producers, and regulators
- New and stronger markets for alternatively grown and processed farm products

Despite the fact that farmers often suffer a reduction in both yield and profit in the first year or two of the transition period, most of those that persist eventually realize both economic and ecological benefits from having made the conversion (Swezey et al., 1994; Gliessman et al., 1996). Part of the success of the transition is based on a farmer's ability to adjust the economics of the farm operation to the new relationships of farming with a different set of input and management costs, as well as adjusting to different market systems and prices.

The conversion to ecologically based agroecosystem management results in an array of ecological changes in the system (Gliessman, 1986). As the use of synthetic agrochemicals is reduced or eliminated, and nutrients and biomass are recycled within the system, agroecosystem structure and function change as well (Jansen et al., 1995). A range of processes and relationships are transformed, beginning with aspects of basic soil structure, organic matter content, and diversity and activity of soil biota (Dick et al., 1994). Eventually, major changes also occur in the relationships among weed, insect, and disease populations, and in the balance between beneficial and pest organisms. Ultimately, nutrient dynamics and cycling, energy use efficiency, and overall system productivity are impacted (Giampietro et al., 1994). Measuring and monitoring these changes during the conversion period helps the farmer evaluate the success of the conversion process, and provides a framework for determining the requirements for sustainability.

The conversion process can be complex, requiring changes in field practices, day to day management of the farming operation, planning, marketing, and philosophy. The following principles can serve as general guidelines for navigating the overall transformation:

- Shift from throughflow nutrient management to recycling of nutrients, with increased dependence on natural processes such as biological nitrogen fixation and mycorrhizal relationships
- Use renewable sources of energy instead of non-renewable sources
- Eliminate the use of nonrenewable off farm human inputs that have the potential to harm the environment or the health of farmers, farm workers, or consumers
- When materials must be added to the system, use naturally occurring materials instead of synthetic, manufactured inputs
- Manage pests, diseases, and weeds instead of "controlling" them
- Re-establish the biological relationships that can occur naturally on the farm instead of reducing and simplifying them
- Make more appropriate matches between cropping patterns and the productive potential and physical limitations of the farm landscape
- Use a strategy of adapting the biological and genetic potential of agricultural plant and animal species to the ecological conditions of the farm rather than modifying the farm to meet the needs of the crops and animals
- Stress the overall health of the agroecosystem rather than the outcome of a particular crop system or season
- Emphasize conservation of soil, water, energy, and biological resources
- Incorporate the idea of long-term sustainability into overall agroecosystem design and management

The integration of these principles creates a synergism of interactions and relationships on the farm that eventually leads to the development of the properties of sustainable agroecosystems listed in Table 1.1. Emphasis on particular principles will vary, but all of them can contribute greatly to the conversion process.

## 1.4 ESTABLISHING CRITERIA FOR AGRICULTURAL SUSTAINABILITY

If we are concerned about maintaining the productivity of our food production systems over the long-term, we need to be able to distinguish between systems that remain temporarily productive because of their high levels of inputs, and those that remain productive indefinitely. This involves being able to predict where a system is headed — how its productivity will change in the future. We can do this through analysis of today's agroecosystem processes and conditions.

The central question involves how a system's ecological parameters are changing over time. Are the ecological foundations of system productivity being maintained or enhanced, or are they being degraded in some way? An agroecosystem that will someday become unproductive gives us numerous hints of its future condition. Despite continuing to give acceptable yields, its underlying foundation is being destroyed. Its topsoil may be gradually eroding year by year; salts may be accumulating; the diversity of its soil biota may be declining. Inputs of fertilizers and pesticides may mask these signs of degradation, but they are there nonetheless for the farmer or agroecological researcher to detect. In contrast, a sustainable agroecosystem will show no signs of underlying degradation. Its topsoil depth will hold steady or increase; the diversity of its soil biota will remain consistently high.

In practice, however, distinguishing between systems that are degrading their foundations and those that are not is not as straightforward as it may seem. A multitude of interacting ecological parameters, determine sustainability — considering each one independently or relying on only a few may prove misleading. Moreover, some parameters are more critical than others, and gains in one area may compensate for losses in another. A challenge for agroecological research is to learn how the parameters interact and to determine their relative importance (Gliessman, 1984, 1987, 1995; Edwards, 1987).

Analysis of agroecosystem sustainability or unsustainability can be applied in a variety of ways. Researchers or farmers may want to do any of the following, alone or in combination:

- Provide evidence of unsustainability on an individual farm in order to stimulate changes in the practices on that farm
- Provide evidence of the unsustainability of conventional practices or systems to argue for changes in agricultural policy and societal values regarding agriculture
- Predict how long a system can remain productive
- Prescribe specific ways to avert productive collapse short of complete redesign of the agroecosystem
- Prescribe ways to convert to a sustainable path through complete agroecosystem redesign
- Suggest ways to restore or regenerate a degraded agroecosystem

Although these applications of sustainability analysis overlap, each represents a different focus and requires a different type of research approach.

## 1.4.1  The Productivity Index

One important aspect of sustainability analysis is to use a wholistic basis for analyzing an agroecosystem's most basic process — the production of biomass. Conventional agriculture is concerned with this process in terms of yield. How the harvest output is created is not important as long as the production is as high as possible. For sustainable agroecosystems, however, measurement of production alone is not adequate because the goal is sustainable production. Attention must be paid to the processes that enable production. This means focusing on productivity — the set of processes and structures actively chosen and maintained by the farmer to produce the harvest.

From an ecological perspective, productivity is a process in ecosystems that involves the capture of light energy and its transformation into biomass. Ultimately, it is this biomass that supports the processes of sustainable production. In a sustainable agroecosystem, therefore, the goal is to optimize the process of productivity so as to ensure the highest yield possible without causing environmental degradation, rather than to strive for maximum yields at all costs. If the processes of productivity are ecologically sound, sustainable production will follow.

One way of quantifying productivity is to measure the amount of biomass invested in the harvested product in relation to the total amount of standing biomass present in the rest of the system. This is done through the use of the productivity index, represented by the following formula:

$$\text{Productivity Index (PI)} = \frac{\text{Total biomass accumulated in the system}}{\text{Net primary productivity (NPP)}}$$

The productivity index provides a way of measuring the potential for an agroecosystem to sustainably produce a harvestable yield. It can be a valuable tool in both the design and the evaluation of sustainable agroecosystems. A PI value can be used as an indicator of sustainability if we assume that there is a positive correlation between the return of biomass to an agroecosystem and the system's ability to provide harvestable yield.

The value of the productivity index will vary between a low of 1 for the most extractive annual cropping system, to a high of about 50 in some natural ecosystems, especially ecosystems in the early stages of succession. The higher the PI of a system, the greater its ability to maintain a certain harvest output. For an intensive annual cropping system, the threshold value for sustainability is 2. At this level, the amount of biomass returned to the system each season is equal to what is removed as yield, which is the same as saying that half of the biomass produced during the season is harvested, and half returned to the system.

NPP does not vary much between system types (it ranges from 0 to 30 t/ha/yr); what varies from system to system is standing biomass (it ranges from 0 and 800 t/ha). When a larger portion of NPP is allowed to accumulate as biomass or standing crop, the PI increases as does the ability to harvest biomass without compromising sustainable system functioning. One way of increasing the standing biomass of the system is to combine annuals and perennials in some alternating pattern in time and space.

To be able to apply the PI in the most useful manner, we must find answers to a number of questions: How can higher ratios be sustained over time? How is the ratio of the return of biomass to the amount of biomass harvested connected to the process of productivity? What is the relationship between standing crop or biomass in an agroecosystem and the ability to remove biomass as harvest or yield?

## 1.4.2  Ecological Conditions of Sustainable Function

The science of ecology provides us with a set of ecological parameters that can be studied and monitored over time to assess movement toward or away from sustainability (Gliessman, 1998a, 2000; Stinner and House, 1987). These parameters include species diversity, organic matter content of the soil, and topsoil depth. For each parameter, agroecological theory suggests a general type of condition or quality that is necessary for sustainable functioning of the system — such as high diversity, high organic matter content, and thick topsoil. The specific rates, levels, values, and statuses of these parameters that together indicate a condition of sustainability will vary for each agroecosystem because of differences in farm type, resources used, local climate, and other site-specific variables. Each system must be studied separately to generate sets of system specific indicators of sustainability.

The parameters listed in Table 1.2 provide a framework for research focusing on what is required for sustainable function of an agroecosystem. Explanations of the role of each parameter in a sustainable system are not provided here, but other chapters in this volume discuss many of them in greater detail.

**Table 1.2  Parameters Related to Agroecosystem Sustainability**

### A. Characteristics of the Soil Resource

*Over the long-term*
  a. soil depth, especially that of the topsoil and the organic horizon
  b. percent of organic matter content in the topsoil and its quality
  c. bulk density and other measures of compaction in the plow zone
  d. water infiltration and percolation rates
  e. salinity and mineral levels
  f. cation-exchange capacity and pH
  g. ratios of nutrient levels, particularly C:N

*Over the short term*
  h. annual erosion rates
  i. efficiency of nutrient uptake
  j. availability and sources of essential nutrients

### B. Hydrogeological Factors

*On-farm water use efficiency*
  a. infiltration rates of irrigation water or precipitation
  b. soil moisture-holding capacity
  c. rates of erosional losses
  d. amount of waterlogging, especially in the root zone
  e. drainage effectiveness
  f. distribution of soil moisture in relation to plant needs

*Surface water flow*
  g. sedimentation of water courses and nearby wetlands
  h. agrochemical levels and transport
  i. surface erosion rates and gully formation
  j. effectiveness of conservation systems in reducing non-point-source pollution

*Ground water quality*
  k. water movement downward into the soil profile
  l. leaching of nutrients, especially nitrates
  m. leaching of pesticides and other contaminants

*continued*

**Table 1.2 (continued)   Parameters Related to Agroecosystem Sustainability**

### C. Biotic Factors

*In the Soil*

a. total microbial biomass in the soil
b. rates of biomass turnover
c. diversity of soil microorganisms
d. nutrient cycling rates in relation to microbial activity
e. amounts of nutrients or biomass stored in different agroecosystem pools
f. balance of beneficial to pathogenic microorganisms
g. rhizosphere structure and function

*Above the Soil*

h. diversity and abundance of pest populations
i. degree of resistance to pesticides
j. diversity and abundance of natural enemies and beneficials
k. niche diversity and overlap
l. durability of control strategies
m. diversity and abundance of native plants and animals

### D. Ecosystem-level Characteristics

a. annual production output
b. components of the productivity process
c. diversity: structural, functional, vertical, horizontal, temporal
d. stability and resistance to change
e. resilience and recovery from disturbance
f. intensity and origins of external inputs
g. sources of energy and efficiency of use
h. nutrient cycling efficiency and rates
i. population growth rates
j. community complexity and interactions

### E. Ecological Economics (Farm Profitability)

a. per unit production costs and returns
b. rate of investment in tangible assets and conservation
c. debt loads and interest rates
d. variance of economic returns over time
e. reliance on subsidized inputs or price supports
f. relative net return to ecologically based practices and investments
g. off-farm externalities and costs that result from farming practices
h. income stability and farming practice diversity

### F. The Social and Cultural Environment

a. equitability of return to farmer, farm laborer, and consumer
b. autonomy and level of dependence on external forces
c. self-sufficiency and the use of local resources
d. social justice, especially cross-cultural and intergenerational
e. equitability of involvement in the production process

# REFERENCES

Altieri, M.A., Why study traditional agriculture? in Carroll, C.R., Vandermeer, J.H., and Rosset, P.M., Eds., *Agroecology*, McGraw-Hill, New York, 1990, 551–564.

Altieri, M.A., *Agroecology: the Science of Sustainable Agriculture*, 2nd ed., Westview Press, Boulder, CO, 1995.

Conway, G.R., Agroecosystem analysis, *Agricultural Adm.*, 20, 31–55, 1985.

Dick, R.P., Soil enzyme activities as indicators of soil quality, in Doran, J.W., Coleman, D.C., Bezsicek, D.F., and Stewart, B.A., Eds., *Defining Soil Quality for a Sustainable Environment*, Special Publication 35, Soil Science Society of America, Madison, WI, 1994, 107–124.

Edwards, C.A., The concept of integrated systems in lower input/sustainable agriculture, *Amer. J. Alternative Agric.*, 2, 148–152, 1987.

Edwards, C.A., Lal, R., Madden, P., Miller, R.H., and House, G., *Sustainable Agricultural Systems*, Soil and Water Conservation Society, Ankeny, IA, 1990.

Giampietro, M., Bukkens, S.G.F., and Pimentel, D., Models of energy analysis to assess the performance of food systems, *Agricultural Systems*, 45, 19–41, 1994.

Gliessman, S.R., An agroecological approach to sustainable agriculture, in Jackson, W., Berry, W., and Colman, B., Eds., *Meeting the Expectations of the Land*, Northpoint Press, Berkeley, CA, 1984, 160–171.

Gliessman, S.R., The ecological element in farm management, in *Proceedings of a Conference on Sustainability of California Agriculture*, University of California, Davis, CA, 1986.

Gliessman, S.R., Species interactions and community ecology in low external-input agriculture, *Am. J. Alternative Agric.*, 11, 160–165, 1987.

Gliessman, S.R., Ed., *Agroecology: Researching the Ecological Basis for Sustainable Agriculture*, Springer-Verlag Series in Ecological Studies, Springer-Verlag, New York, 78, 1990.

Gliessman, S.R., Sustainable agriculture: an agroecological perspective, in Andrews, J.S. and Tommerup, I.C., Eds., *Advances in Plant Pathology*, 11, 45–56, 1995.

Gliessman, S.R., Agroecology: researching the ecological processes in sustainable agriculture, in Chou, C.H. and Shan, K.T., Eds., *Frontiers in Biology: The Challenges of Biodiversity, Biotechnology, and Sustainable Agriculture*, Academia Sinica, Taipei, Taiwan, 1998, 173–186.

Gliessman, S.R., *Agroecology: Ecological Processes in Sustainable Agriculture*, Lewis Publishers, Boca Raton, FL, 2000.

Gliessman, S.R., Werner, M.R., Swezey, S., Caswell, E., Cochran, J., and Rosado-May, F., Conversion to organic strawberry management: changes in ecological processes, *Calif. Agric.*, 50, 24–31, 1996.

Jansen, D.M., Stoorvogel, J.J., and Schipper, R.A., Using sustainability indicators in agricultural land use analysis: an example from Costa Rica, *Netherlands J. Agric. Sci.*, 43, 61–82, 1995.

Klee, G., *World Systems of Traditional Resource Management*, Halstead, New York, 1980.

Moldan, B., Billharz, S., and Matravers, R., Sustainability indicators: a report on the project on indicators of sustainable development, John Wiley & Sons, Scope Rep. 58, Chichester, 1997.

Munasinghe, M. and Shearer, W., Eds., Defining and measuring sustainability: the biophysical foundations, World Bank, Washington, D.C., 1995.

National Research Council (NRC), *Alternative Agriculture*, National Academy Press, Washington, D.C., 1989.

Odum, E.P., Properties of agroecosystems, in Lowrance, R., Stinner, B.R., and House, G.J., Eds., *Agricutural Ecosystems: Unifying Concepts*, John Wiley & Sons, New York, 5–12, 1984.

Organization for Economic Cooperation and Development, *Towards Sustainable Development: Environmental Indicators*, OECD, Washington, D.C., 1998.

Stinner, B.R. and House, G.J., Role of ecology in lower-input, sustainable agriculture: an introduction, *Am. J. Alternative Agric.*, 2, 146–147, 1989.

Swezey, S.L., Rider, J., Werner, M.W., Buchanan, M., Allison, J., and Gliessman, S.R., Granny Smith conversions to organic show early success, *Calif. Agric.*, 48, 36–44, 1994.

Wilken, G.C., *Good Farmers: Traditional Agricultural Resource Management in Mexico and Central America*, University of California Press, Berkeley, 1988.

# CHAPTER 2

# Increasing Sustainability in Mediterranean Cropping Systems with Self-Reseeding Annual Legumes

Fabio Caporali and Enio Campiglia

## CONTENTS

## 2.1 INTRODUCTION

Modern specialized agricultural systems carry out productive functions only by adding large auxiliary energy inputs (synthetic fertilizers, pesticides, etc.) and paying little attention to environmental degradation and human health risks. To correct these negative tendencies it is necessary to return to environmentally sound agriculture and to implement it in modern agroecosystems (Caporali et al., 1989; Marsh, 1997).

The environmental soundness of an agroecosystem is closely linked to its complexity: complex agricultural systems are regarded as more dependable in production and more sustainable in terms of resource conservation than simple ones (Stinner et al., 1997; Vandermeer et al., 1998). A key aspect of agroecosystem complexity is cropping diversity, which can be increased both spatially (e.g., through intercropping) and temporally (through crop rotations). By developing agroecosystems with greater crop diversity, we come closer to imitating the more complicated structures and functions of natural communities, in which physical and biological resources are maximally utilized and integrated.

In search of strategies for increasing sustainability in cropping systems through increased cropping diversity, we have focused for 10 years on the use of alternative plant resources, such as the self-reseeding winter annual legumes (*Trifolium* and *Medicago* spp.) native to the Mediterranean environment. Although annual self-reseeding legumes are well known forage crops in cereal-ley farming under the Mediterranean climate throughout the world, their use in cash crop sequences is virtually unknown. Nevertheless, they have many valuable traits that can be exploited in cash crop sequences, as they (a) grow during the cool season; (b) die in the early summer; (c) regenerate after fall rains providing cover that can be used as either a green manure or a dry mulch for the succeeding crop; (d) tolerate shade; (e) provide weed control through good growth coverage; (f) provide significant quantities of fixed N while conserving soil and water resources and sustaining or improving soil productivity; and (g) allow the use of minimum tillage or no till practices.

Because of these characteristics and advantages, winter growing legumes might be used for improving the agroecological performance of a conventional cash crop sequence such as the 2-year rotation between a winter cereal (wheat or barley) and a summer crop (sunflower or maize), the most common cropping pattern followed in the arable land of central Italy. This chapter describes our research over 10 years, starting from the screening of the self-reseeding legume species and cultivars and ending with the implementation and performance assessment of the entire alternative cropping system.

## 2.2 THE IMPACTS OF CONVENTIONAL ROTATIONS IN CENTRAL ITALY

In the arable hilly land of central Italy, the most usual cash crop sequence is the 2-year rotation between a rain fed winter cereal (wheat or barley) and a summer crop (rain fed sunflower or irrigated maize). This rotation, which involves the application of N fertilizers, chemical weeding, and frequent tillage (Caporali and Onnis, 1991), is energy intensive, costly, and environmentally harmful.

Since the summer crop is usually grown following a tilled fallow period, there is a serious risk of loss of soil through erosion, loss of organic matter through mineralization, and leaching of nitrates into ground and surface water. The common practice is to plow the soil during the summer months and leave it bare until mid autumn, when winter cereals are sown and nitrogen fertilizer is applied, or until early spring, when the summer crop is sown. The lack of soil cover during autumn, or autumn and/or winter, the period of heaviest rainfall, allows both mineralization of the soil and N leaching. It is well documented that nitrate levels in the surface and ground water of agricultural watersheds in central Italy increase during the winter period (Caporali et al., 1981; Nannipieri et al., 1985); this is partially due to the widespread use of conventional rotation.

## 2.3  CONCEIVING AN ALTERNATIVE CROPPING SYSTEM

In the Mediterranean environment, legumes have evolved well adapted biological forms (therophytes) that are able to grow during the moist, cold season and set fruits before the dry, hot season becoming seeds on or in the ground. As they are able to regenerate after autumn rainfall, when a new life cycle starts, self-reseeding annual legumes are annuals but behave practically like polyannuals.

The legume life cycle meshes well with the conventional 2-year rotation described above. In the alternative cropping system that we have conceived, the annual legumes grow as a living mulch in the winter cereal; then, after reseeding and emergence with the autumn rains, they grow through the next autumn and winter, becoming either a green manure or a dry mulch for the succeeding summer crop (see Figure 2.1). During the intercropped cereal phase of the rotation, the annual legumes do not compete with the winter crop for water because rainfall is typically abundant during this period.

This alternative cropping system has the potential to induce a significant shift toward a less energy intensive and a more environmentally friendly type of management, while maintaining the traditional sequence of cash crops and providing more innovative and flexible patterns of cover cropping (Caporali et al., 1993).

## 2.4  SCREENING OF LEGUME SPECIES AND CULTIVARS

For the alternative cropping system to function successfully, the legume needs to meet three main requirements: (a) perform as a living mulch in winter cereals;

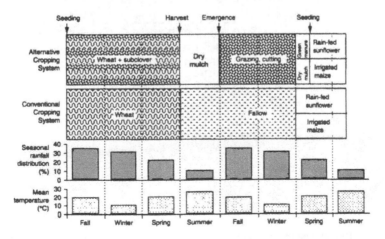

**Figure 2.1**  Cropping pattern comparison showing precipitation and temperature regime at Viterbo (Central Italy).

(b) regenerate abundantly after the cereal harvest; and (c) cover the ground during winter to furnish sufficient biomass for use as a green manure or a dry mulch for the succeeding summer crop. We conducted many trials to compare the abilities of legumes to meet these requirements.

In an initial set of trials we used barley as the cereal component of a same intercropping pattern, with many different winter annual legumes as living mulches. The trials were implemented according to the replacement series methodology (Willey, 1979), which compares same density combinations ranging from a pure stand of one component through various mixtures to a pure stand of the other component. In our case, a single mixture was used, consisting of barley and legume in a 50:50 proportion (see Figure 2.2). To create the sole legume and sole barley stands, we applied seed at a recommended rate to achieve a density level of 400 plants/m$^2$ for the barley and 300 plants/m$^2$ for the legume; to create mixed stands, we applied the seeds of both components at half the recommended rate. No nitrogen fertilizer was applied during the whole crop cycle and no weeding was necessary due to poor weed development in both the single and mixed crops.

As a measure of the biological efficiency of the intercropping systems, we used the LER, which is the index most frequently used by researchers in multiple cropping (Francis, 1989). Legume seedlings were counted in all the mixed crops after the autumn emergence in both seeding and self-reseeding conditions in permanent quadrates (0.25 m$^2$), and a re-establishment ratio was calculated. Plant material was dried at 70°C until constant weight; total nitrogen was determined by the Kjeldahl method.

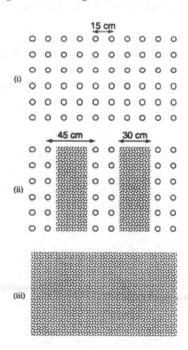

**Figure 2.2**  Row and band arrangement in the sod strip intercropping system. (i) = barley (400 plants/m$^2$); (ii) = barley (200 p/m$^2$) + winter legume (150 p/m$^2$); (iii) = broadcast sown legume (300 p/m$^2$). Barley plants = o, legumes = shaded areas.

Table 2.1   Barley Grain Yield (DM), Barley Partial LER, and Aboveground Biomass Total
LER of Barley Intercropped with Different Self-Reseeding Annual Legumes

| Cropping System | Barley Yield (DM) (kg/ha) | Partial LER for Barley | Total LER (Aboveground Biomass) |
|---|---|---|---|
| Barley + *T. subterraneum* Woogenellup | 4495 a | 1.20 a | 1.29 bc |
| Barley + *T. subterraneum* Mount Barker | 4283 ab | 1.14ab | 1.53 a |
| Barley + *T. subterraneum* Seaton Park | 4280 ab | 1.14 ab | 1.22 bc |
| Barley + *T. subterraneum* Daliak | 3957 abcd | 1.06 abc | 1.17 bc |
| Barley + *T. yanninicum* Meteora | 3949 abcd | 1.05 abc | 1.36 ab |
| Barley + *T. brachycalycinum* Clare | 3861 bcde | 1.03 abc | 1.60 a |
| Barley + *M. polymorpha* Serena | 3825 bcdef | 1.02 abc | 1.10 bc |
| Barley sole crop | 3744 bcdefg | — | — |
| Barley + *M. polymorpha* Circle Valley | 3553 cdefgh | 0.95 bcd | 1.01 c |
| Barley + *T. subterraneum* Dalkeit | 3446 defghi | 0.92 cd | 1.20 bc |
| Barley + *T. yanninicum* Larisa | 3426 defghi | 0.91 cd | 1.37 ab |
| Barley + *T. hirtum* Kikon | 3271 efghi | 0.87 cde | 1.02 c |
| Barley + *M. rugosa* Paraponto | 3240 fghij | 0.86 cde | 1.12 bc |
| Barley + *T. subterraneum* Junee | 3141 ghij | 0.84 cde | 1.09 bc |
| Barley + *M. truncatula* Parabinga | 3113 hij | 0.83 cde | 1.13 bc |
| Barley + *M. rugosa* Sapo | 2841 ijk | 0.76 def | 0.99 c |
| Barley + *M. truncatula* Paraggio | 2651 jkl | 0.71 def | 1.02 c |
| Barley + *T. michelianum* Giorgia | 2500 kl | 0.67 ef | 1.12 bc |
| Barley + *T. brachycalycinum* Altura | 2229 l | 0.59 f | 1.17 bc |

Values within columns followed by the same letter do not differ significantly at the 5% probability level according to a Fisher LSD Protected Test.

Table 2.1 reports grain yield and LER values of the intercropped components. The biological efficiency of the intercropping system, expressed as the ability to incorporate solar energy into biomass, was generally higher than or equal to that of the sole crop system (total LER > 1); particularly, it was significantly higher when *T. brachycalicyinum* cv. Clare, *T. subterraneum* cv. Mount Barker and *T. yanninicum* cv. Meteora were intercropped to barley. For these legumes, the LER values of the intercrop were 1.60, 1.53, and 1.36, respectively. In this kind of intercropping, where only the cereal component is to be harvested and the legume component is to be left on the ground to regenerate, the most successful mixtures are those that combine high total LER and high partial LER for the grain. This is the case for the mixture of barley and *T. subterraneum* cv. Mount Barker, which produces 64% more grain than sole barley and 11% less biomass than subclover alone.

Table 2.2 reports the components of barley grain yield under the different cropping system conditions. The number of fertile culms/m² was the yield character of barley most influenced in the intercrop. We conclude that barley productivity depends mostly on factors influencing its tillering capacity. *T. subterraneum* cv. Mount Barker was the only legume to induce the cereal to produce a statistically significant increase in fertile culms compared to the cereal sole crop.

Data in Table 2.3 show the re-establishment ability of the intercropped legumes in the first year after reseeding. Autumn emergence after reseeding ranged on the whole from 0 to 310 seedlings/m². The *Medicago* species and the *Trifolium* cultivars *T. hirtum* and *T. michelianum* did not show any regenerating ability. Seedling density

**Table 2.2   Yield Characters of Barley Intercropped with Different Self-Reseeding Annual Legumes**

| Cropping System | Culms/m$^2$ | Dry Mass per 1000 Grains (g) | Grains/Ear |
|---|---|---|---|
| Barley + *T. subterraneum* Mount Barker | 525 a | 37.18 ab | 28.0 abc |
| Barley + *T. subterraneum* Woogenellup | 462 b | 38.26 a | 28.3 abc |
| Barley + *M. polymorpha* Circle Valley | 436 bc | 31.99 cde | 29.7 abb |
| Barley + *T. subterraneum* Seaton Park | 433 bc | 33.53 abcd | 27.0 abc |
| Barley + *T. subterraneum* Dalkeit | 427 bc | 28.27 e | 29.3 abb |
| Barley + *M. polymorpha* Serena | 426 bc | 33.60 abcd | 28.7 abc |
| Barley sole crop | 411 bc | 34.73 abc | 28.7 abc |
| Barley + *T. yanninicum* Meteora | 408 bc | 30.86 cde | 29.0 abc |
| Barley + *T. hirtum* Kikon | 401 cd | 32.57 bcde | 24.7 cd |
| Barley + *T. subterraneum* Daliak | 396 cde | 31.99 cde | 30.7 a |
| Barley + *T. brachycalycinum* Clare | 396 cde | 29.78 cde | 27.3 abc |
| Barley + *M. truncatula* Parabinga | 395 cde | 31.49 cde | 28.3 abc |
| Barley + *T. subterraneum* Junee | 385 cdef | 32.42 bcde | 28.7 abc |
| Barley + *M. rugosa* Paraponto | 346 defg | 34.70 abc | 21.0 d |
| Barley + *T. brachycalycinum* Altura | 339 efg | 28.58 de | 26.3 abc |
| Barley + *T. yanninicum* Larisa | 337 efg | 34.10 abc | 30.7 a |
| Barley + *T. michelianum* Giorgia | 329 fg | 31.17 cde | 26.3 abc |
| Barley + *M. truncatula* Paraggio | 324 g | 33.41 abcd | 25.3 bcd |
| Barley + *M. rugosa* Sapo | 307 g | 30.33 cde | 29.3 ab |

Values within a column followed by the same letter do not differ significantly at the 5% probability level according to a Fisher LSD Protected Test.

in *T. subterraneum* cultivars ranged from 145 to 310 seedlings/m$^2$; that range of values is considered appropriate for stand establishment as a winter cover crop (Evers et al., 1988). Particularly notable was the performance of *T. subterraneum* cv. Mount Barker, which showed a re-establishment ratio of 1.20 (310/258 seedlings/m$^2$).

The above mentioned results suggest that the establishment of winter annual legumes as living mulches is practical in barley in the form of sod-strip intercropping. Several of the tested *T. subterraneum* subspecies and cultivars were able to grow sufficiently when intercropped, without reducing, and in some cases improving, barley grain yield performance, and to regenerate successfully. Cultivars selected on the basis of the best performances are listed in Table 2.4. Most of our further research on the alternative cropping system was conducted by using the top performing *T. subterraneum* cv Mount Barker.

## 2.5 PRACTICAL PERFORMANCE OF THE ALTERNATIVE CROPPING SYSTEM

### 2.5.1   Problems with the Winter Crop Component

Because the seed rate of the intercropped cereal is exactly half of that in the pure stand, it is necessary to rely on a cereal genotype that possesses strong tillering capacity. Unfortunately, modern breeding trends are oriented toward creating non-tillered or uniculm varieties, suitable for cereal growth and yield in a pure stand,

Table 2.3   Number of Seedlings Emerged and Reestablishment Ratio of the Self-
Reseeding Annual Legumes

| Species | Seedlings/m² | | Re-est. Ratio (S2/S1) |
|---|---|---|---|
| | 1988 Emergence (S1) | 1989 Emergence (S2) | |
| Barley + T. subterraneum Mount Barker | 258 | 310 | 1.20 (0.34 a) |
| Barley + T. subterraneum Junee | 250 | 285 | 1.15 (0.33 a) |
| Barley + T. yanninicum Larisa | 283 | 289 | 1.02 (0.31 ab) |
| Barley + T. brachycalycinum Altura | 325 | 262 | 0.80 (0.26 bc) |
| Barley + T. subterraneum Dalkeit | 345 | 243 | 0.71 (0.23 cd) |
| Barley + T. yanninicum Meteora | 358 | 243 | 0.68 (0.23cd) |
| Barley + T. brachycalycinum Clare | 358 | 242 | 0.67 (0.22 cd) |
| Barley + T. subterraneum Seaton Park | 262 | 145 | 0.56 (0.19 de) |
| Barley + T. subterraneum Woogenellup | 325 | 175 | 0.54 (0.19 de) |
| Barley + T. subterraneum Daliak | 383 | 158 | 0.41 (0.15 e) |
| Barley + T. michelianum Giorgia | 458 | 48 | 0.11 (0.05 f) |
| Barley + M. rugosa Sapo | 300 | 7 | 0.02 (0.01 f) |
| Barley + T. hirtum Kikon | 225 | 4 | 0.02 (0.01 f) |
| Barley + M. rugosa Paraponto | 217 | 4 | 0.02 (0.01 f) |
| Barley + M. polymorpha Serena | 267 | 3 | 0.01 (0.00 f) |
| Barley + M. truncatula Paraggio | 283 | 0 | 0.00 (0.00 f) |
| Barley + M. truncatula Parabinga | 300 | 0 | 0.00 (0.00 f) |
| Barley + M. polymorpha Circle Valley | 250 | 0 | 0.00 (0.00 f) |

Figures in parentheses are the Log Transformed Values to be considered for mean comparison. Values followed by the same letter do not differ significantly at the 5% probability level
according to a Fisher LSD Protected Test.

depriving both intercropping research and practice of their necessary genotype basis.
This fact was confirmed by our research comparing the conventional and the alternative cropping system using wheat as the winter cereal component. Although we
adopted one of the best performing modern winter wheat cultivars for tillering
capacity (cv Pandas), yield measures were still constrained by the tillering capacity
of the cereal, as they were with barley.

Tillering capacity of modern varieties is affected significantly by nitrogen availability. As indicated by the data reported in Table 2.5, the number of fertile culms
is the main factor influencing wheat grain yield; the number of fertile culms increases
with the presence of N fertilizer. The heavy N leaching that occurs in wet winters
can greatly reduce both tillering capacity and yield of intercropped wheat when
compared to the pure stand. It is likely that a reduced tillering capacity is associated
with a less developed root system which does not allow a good uptake of water and

Table 2.4  Synopsis of the Self-Reseeding Annual Clovers That Perform Best as Living Mulches In Barley

| Cropping System | Barley | | | | Legume | |
|---|---|---|---|---|---|---|
| | Total LER | Partial LER | Absolute Yield | Aboveground Biomass | N Yield | Re-est. Ratio |
| Barley + *T. subterraneum* Mount Barker | + | + | + | + | + | + |
| Barley + *T. subterraneum* Woogenellup | 0 | + | + | 0 – | 0 – | 0 |
| Barley + *T. subterraneum* Seaton Park | 0 | + | + | 0 – | 0 – | 0 |
| Barley + *T. subterraneum* Daliak | 0 | 0 | + | 0 – | 0 – | 0 |
| Barley + *T. yanninicum* Meteora | + | 0 | + | 0 | 0 | 0 |
| Barley + *T. brachycalycinum* Clare | + | 0 | + 0 | + | + | 0 |

Performance Level: High (+); Medium (0); Low (–).

Table 2.5  Mean Effect of Cropping System, N Fertilizer Application, and Type of Weeding on Grain Yield and Yield Components of Wheat in Two Different Growing Seasons

| | 1989–90 (drier; total rainfall 287 mm) | | | | 1990–91 (wetter; total rainfall 489 mm) | | | |
|---|---|---|---|---|---|---|---|---|
| | Grain Yield (kg/ha, DM) | Fertile culms per m² | Grains per ear | Mass per 1000 grains (g) | Grain Yield (kg/ha, DM) | Fertile culms per m² | Grains per ear | Mass per 1000 grains (g) |
| Wheat + Subclover | 3582 a | 331 a | 31.9 a | 54.0 a | 4164 a | 287 a | 35.2 a | 51.9 a |
| Wheat (conventional) | 3601 a | 396 b | 29.7 a | 48.9 b | 4702 b | 405 b | 31.4 a | 47.6 a |
| N 0 kg/ha | 3135 a | 333 a | 29.8 a | 50.3 a | 3270 a | 267 a | 31.9 a | 49.2 a |
| N 130 kg/ha | 4042 b | 373 b | 32.5 a | 54.5 a | 5418 b | 387 b | 36.1 a | 51.7 a |
| Hand weeded | 3647 a | 353 a | 30.3 a | 55.4 a | 4812 a | 343 a | 34.3 a | 51.4 a |
| Unweeded | 3530 a | 353 a | 32.0 a | 49.3 a | 3876 b | 311 b | 33.6 a | 49.6 a |

For each treatment, values within columns followed by the same letter do not differ significantly at the 5% probability level according to a Fisher protected test.
Modified after Caporali and Campiglia, 1993.

nutrients in the soil. The presence of weeds that compete with wheat for nitrogen can also reduce significantly the number of wheat culms, especially in wet years when weed density is usually higher.

The intercropped wheat yielded as much as the sole wheat in the drier and less productive year while a higher number of grains per ear and a higher grain mass counterbalanced the lower number of fertile culms. This result agrees with the law of constant final yield, which states that over a wide range of population densities, yield/unit area becomes independent of the number of plants sown (Harper, 1977). However, a yield decrease of nearly 12% was recorded for the intercropped wheat in the wetter and more productive year when total rainfall during the wheat cycle amounted to 489 mm. As the water availability improved, tillering capacity seemed to become a major constraint to obtaining higher grain yield in the intercrop.

## 2.5.2  Subclover as a Green Manure for Sunflower

Using the subclover as a green manure significantly improved the performance of sunflowers, one of the commonly planted summer crops in the conventional rotation. The green manuring improved overall yield, reduced the effect of drought, and depressed weed populations.

The grain yield of sunflowers grown conventionally (no green manuring and N fertilizer applied) during the dry summer of 1991 (no rainfall during the 2 months before sunflower harvest) was half that of sunflowers grown conventionally in the succeeding wetter year (see Table 2.6). Drought is the most important factor limiting sunflower yield performance in a Mediterranean climate like that of central Italy. During the dry growing season, however, the grain yield of sunflowers grown with whole plant green manuring was 29% greater than the grain yield of conventional sunflowers when N fertilizer was added and 25% greater without fertilizer.

The data show that green manuring had a stronger effect on sunflower grain yield than the application of N fertilizer, regardless of rainfall. In the drier year, N fertilization had no effect on grain production at all in the conventionally grown sunflowers. In the wetter growing season, even a partial subclover green manuring (stubble + roots) was so effective that sunflower grain yield in the alternative

Table 2.6  Effect of Subclover Green Manuring on Sunflower Grain Yield (kg/ha DM)

|  | 1991 (drier growing season) | | 1992 (wetter growing season) | |
|---|---|---|---|---|
|  | N 130 kg/ha | N 0 kg/ha | N 130 kg/ha | N 0 kg/ha |
| Subclover total green manure (whole plant) | 1467 | 1391 | 4248 | 2957 |
| Subclover partial green manure (stubble + roots) | 1304 | 1123 | 3467 | 2692 |
| No green manure (conventional cropping system) | 1045 | 1049 | 2325 | 997 |

Modified after Caporali and Campiglia, 1993.

Table 2.7  Linear Regression Equations and Correlation Coefficients Between the Amount of Subclover Biomass (kg/ha DM) (x) Plowed in and the Productive Traits (y) of Sunflower

| Trait | Correlation Coefficient | Regression Equation |
|---|---|---|
| Grain yield (kg/ha DM) | 0.93[a] | y = 819.94 + 0.28 x |
| Seed per head | 0.83[b] | y = 402.54 + 0.05 x |
| Aboveground biomass (kg/ha DM) | 0.84[b] | y = 3403.42 + 1.04 x |
| Plant height (cm) | 0.93[a] | y = 46.91 + 0.01 x |
| Head diameter (cm) | 0.91[b] | y = 10.17 + 0.0007 x |

[a] Significant at the 0.01 probability level
[b] Significant at the 0.05 probability level
Data from plots with no N fertilization and with mechanical weeding.
Modified after Caporali and Campiglia, 1993.

cropping system was higher than that of the conventional one fertilized with 130 kg/ha of inorganic N. A positive correlation between the amount of subclover aboveground biomass plowed in and both vegetative and productive characteristics of sunflower was found for the plots with mechanical weeding and without nitrogen fertilization. This is expressed in the linear regression equations reported in Table 2.7. Subclover aboveground biomass ranged between 2192 kg/ha DM (in the drier year) and 7262 kg/ha DM (in the wetter year).

Weed stand biomass assessed at sunflower harvest time in the unweeded plots (see Table 2.8) was largely depressed by both subclover green manure treatments in the wetter year (1992). Direct and indirect effects due to subclover green manuring could explain such a performance (Dyck and Liebman, 1992). Two of the most abundant species in the weed community, *Amaranthus retroflexus* and *Chenopodium album*, were largely depressed in density by both green manure treatments (Figure 2.3).

### 2.5.3  Subclover as a Dry Mulch for Maize

Another possible role for subclover in the alternative system is to let it develop until late spring and then use it as a dry mulch for a succeeding crop of irrigated maize. Cover crops are usually chemically suppressed before crop planting to avoid competition (Lanini et al., 1989), but since subclover dies back naturally as temperatures rise in late spring, it is not necessary to suppress it chemically before

Table 2.8  Effect of Green Manure Treatments on Weed Biomass (kg/ha DM) at Sunflower Harvest Time

| Year | Total green manure | Partial green manure | No green manure |
|---|---|---|---|
| 1991 | 3969 | 3905 | 4526 |
| 1992 | 3308 | 3286 | 6579 |

(Year x green manure) LSD (0.05) = 1062.
Modified after Caporali and Campiglia, 1993.

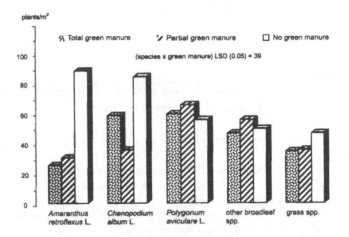

**Figure 2.3**  Effect of green manure treatments on the number of plants per square meter of the most abundant species in the sunflower weed community. (Modified after Caporali and Campiglia, 1993.)

planting the crop. This makes possible a system in which the preceding sod strip intercropping pattern is maintained with each row of maize replacing each pair of wheat rows. A sod strip intercropping system between maize and subclover mulch is generated making maize rows 50 cm apart. Before planting the maize, the subclover stand is mowed and left on the ground. Maize is then direct seeded into the subclover mulch.

Recent results obtained from differently performing subclover mulches on non-fertilized and drip irrigated silage maize are reported in Table 2.9. It is evident that two of the subclover varieties tested have a positive effect on the maize aboveground biomass, suggesting that both a build-up of N and an increase in N mineralization rate occur in the soil as subclover growth increases.

**Table 2.9  The Potential of Subclover as a Dry Mulch for Silage Maize**

| Preceding sod strip intercropping system | Mass of subclover mulch (kg/ha DM) | Maize production (tons/ha DM) | | |
|---|---|---|---|---|
| | | Ear | Stalk | Silage |
| Wheat + *T. subterraneum* Barker | 449.7 a | 10.42 a | 11.12 a | 21.54 a |
| Wheat + *T. subterraneum* Seaton Park | 73.6 c | 7.89 b | 9.44 b | 17.33 b |
| Wheat + *T. brachycalycinum* Clare | 578.9 a | 10.72 a | 11.60 a | 22.32 a |
| Wheat + *T. subterraneum* Daliak | 272.5 b | 8.34 b | 9.36 b | 17.70 b |
| Wheat sole crop | — | 8.34 b | 8.96 b | 17.30 b |

Values within columns followed by the same letter do not differ significantly at the 5% probability level according to a Fisher protected test.

## 2.6 POTENTIAL FOR IMPLEMENTATION AT THE FARM LEVEL

Alternative cropping systems based on the use of self-reseeding annual legumes have the potential to be rapidly adopted in low input farming systems and organic farming systems.

These cropping systems reflect the most important basic principles of organic agriculture. This assessment is based on three major factors:

- Increased use of renewable natural resources and diminished use of fossil fuel-derived resources
- More intensive use of leguminous plants in crop rotations as living mulches, cover crops, and green manures, in order to store more solar energy, conserve soil moisture, and fix atmospheric nitrogen
- More intensive soil coverage by cropping systems to assure a permanent plant canopy during the year and prevent soil degradation

Adopting these alternative cropping systems is a convenient strategy for converting conventional farms into mixed organic farms while maintaining the cash crop sequences most common in a Mediterranean environment. This conversion does not imply, apart from the livestock enterprise, major changes in crop management, farm equipment, or economic outcome.

There is an increasing demand for innovative cropping systems that are seen as more sustainable and which facilitate the conversion process. This is especially true in Italy, due to both the favorable cultural framework and ongoing regulations in Europe (European Union Regulation N. 2092/91) that press farmers to adopt more sustainable methods of production. Currently, Italy ranks first in Europe for both number and total size of organic farms.

Nevertheless, several factors may discourage farmers from using these innovative cropping systems. These include

- A lack of knowledge of self-reseeding annual legumes in crop rotations
- A reluctance among stakeholders (farmers and/or advisors) to adopt intercropping systems in place of the conventional systems of pure crops
- A lack of winter cereal genotypes with high tillering capacity suitable for use in the replacement series intercropping patterns

At present, the alternative cropping systems described in this chapter have been implemented in several commercial farms located in central and southern Italy. One of these farms has been monitored as an agroecosystem since 1995 (Barberi et al., 1998) using an input/output methodology to evaluate energy and financial flows (Caporali et al., 1989). The economic effects of the introduction of the alternative cropping system on this farm will soon be assessed.

## REFERENCES

Barberi, P., Caporali, F., Campiglia, E., and Mancinelli, R., Weed community composition in a mixed farming system in Central Italy. Workshop Proceedings, Dronten, The Netherlands, 25–28 May 1998. Landbouwuniversiteit, Wageningen, 79–83.

Caporali, F. and Onnis, A., Validity of rotation as an effective agroecological principle for a sustainable agriculture, *Agric. Ecosyst. Environm.*, 41, 101–113, 1991.

Caporali, F. and Campiglia, E., Sustainable yield performances of sunflower under dry conditions in Central Italy. Proceedings of Fourth International Conference on Desert Development, Mexico City, 25–30 July 1993, 546–550.

Caporali, F., Nannipieri, P., and Pedrazzini, F., Nitrogen contents of streams draining an agricultural and a forested watershed in Central Italy, *J. Environm. Qual.*, 10(1), 72–76, 1981.

Caporali, F., Nannipieri, P., Paoletti, M.G., Onnis, A., and Tomei, P.E., Concepts to sustain a change in farm performance evaluation, *Agric. Ecosyst. Environ.*, 27, 579–595, 1989.

Caporali, F., Campiglia, E., and Paolini, R., Prospects for more sustainable cropping systems in Central Italy based on subterranean clover as a cover crop, XXVII Intern., *Grassland Congr. Proc.*, Rockhampton, Australia, 2197–2198, 1993.

Dick, E. and Liebman, M., Organic N source effect on crop-weed interactions, *IX IFOAM Intern. Scient. Conf.*, 25–30, São Paulo, 1992.

Evers, G.W., Smith, G.R., and Beale, P.E., Subterranean clover reseeding, *Agron J.*, 80, 855–859, 1988.

Francis, C.A., Biological efficiencies in multiple-cropping systems, *Adv. Agron.*, 42, 1–42, 1989.

Harper, J.L., *Population Biology of Plants*, Academic Press, London, 1977.

Lanini, W.T., Pittenger, D.R., Graves, W., Munoz, F., and Agamalian, H.S., Subclovers as living mulches for managing weeds in vegetables, *Calif. Agric.*, 43(6), 25–27, 1989.

Marsh, J.S., The policy approach to sustainable farming systems in the EU, *Agric. Ecosyst. Environ.*, 64, 103–114, 1997.

Nannipieri, P., Caporali, F., and Arcara, P.G., The effect of land use on the nitrogen biogeochemical cycle in Central Italy, in Caldwell, D.E., Brierley, J.A., and Brierley, C.L., Eds., *Planetary Ecology*, Van Nostrand Reinhold, New York, 1985.

Stinner, D.H., Stinner, B.R., and Marssolf, E., Biodiversity as an organizing principle in agroecosystem management: case studies of holistic resource management in the USA, *Agric. Ecosyst. Environ.*, 62, 199–213, 1998.

Vandermeer, J., van Noordwijk, M., Anderson, J., Ong, C., and Perfecto, I., Global change and multi-species agroecosystems: concepts and issues, *Agric. Ecosyst. Environ.*, 67, 1–22, 1998.

Willey, R.W., Intercropping — its importance and research needs, Part I: Competition and yield advantages, *Field Crop Abs.*, 32, 1–10, 1997.

CHAPTER 3

# Manipulating Plant Biodiversity to Enhance Biological Control of Insect Pests: A Case Study of a Northern California Organic Vineyard

Clara I. Nicholls and Miguel A. Altieri

## CONTENTS

## 3.1 INTRODUCTION

The expansion of monoculture in California has resulted in the simplification of the landscape. One effect of this simplification is a decrease in the abundance and activity of the natural enemies of agricultural pests due to the disappearance of habitats providing them with critical food resources and overwintering sites (Corbett and Rosenheim, 1996). Many scientists are concerned that, with accelerating rates of habitat removal, the contribution to pest suppression by biocontrol agents using these habitats will decline further (Fry, 1995; Sotherton, 1984). This will increase insecticide use with consequent negative effects on the sustainability of agroecosystems.

Many researchers have proposed ways of increasing the vegetational diversity of agricultural landscapes to halt or reverse the decline in natural controls; it is known that biological pest suppression is more effective in diverse cropping systems than in monocultures (Andow, 1991; Altieri, 1994). One such method used in vineyards and orchards is to manage the resident floor vegetation or to plant cover crops. This tactic is designed to maintain habitats for natural enemies and thus enhance their populations. Reductions in mite (Flaherty, 1969) and grape leafhopper populations (Daane et al., 1998) have been observed with such plantings, but the observed biological suppression has not been sufficient from an economic point of view (Daane and Costello, 1998).

Most likely the above studies achieved less than adequate biological suppression of pests because they did not maintain enhanced vegetational diversity for a long enough portion of the growing season. The studies were conducted in vineyards with winter cover crops and/or with weedy resident vegetation, which dried early in the season or was mowed or plowed under at the beginning of the growing season, leaving the systems as virtual monocultures by early summer. Based on this observation, we hypothesize natural enemies may need a green cover for habitat and alternative food during the entire growing season. One way to achieve this condition is to sow summer cover crops that bloom early and throughout the season, thus providing a highly consistent, abundant, and well-dispersed alternative food source, as well as microhabitats, for a diverse community of natural enemies.

Another option is the maintenance or planting of vegetation adjacent to crop fields (Thomas et al., 1991; Nentwing et al., 1998). Ideally, such areas provide alternative food and refuge for predators and parasitoids, thereby increasing natural enemy abundance and colonization of neighboring crops (Altieri, 1994; Corbett and Plant, 1993; Coombes and Sotherton, 1984). Researchers found that entomophagous insects depend on hedges, windbreaks, and forests adjacent to crop fields for their continual existence in agricultural areas (Fry, 1995; Wratten, 1988). Several studies indicate that the abundance and diversity of entomophagous insects within a field are dependent on the plant species composition of the surrounding vegetation. They also depend on its spatial extent and arrangement which affect the distance to which natural enemies disperse into the crop (Lewis, 1965; Pollard, 1968).

Much research has been conducted in California on the role of adjacent vegetation on the *Anagrus epos*–grape leafhopper (*Erythroneura elegantula*) complex. The classic study by Doutt and Nakata (1973) determined the role of riparian habitats, and especially of wild blackberry patches, near vineyards in enhancing the effectiveness of *A. epos* in parasitizing the grape leafhopper. Later research by Kido et al. (1984) established that French prunes adjacent to vineyards could also serve as overwintering sites for *A. epos*; Murphy et al. (1996) detected higher leafhopper parasitism in grape vineyards with adjacent prune tree refuges than in vineyards lacking refuges. Corbett and Rosenheim (1996), however, determined that the effect of prune refuges was limited to a few vine rows downwind; *A. epos* exhibited a gradual decline in vineyards with increasing distance from the refuge. This finding indicates an important limitation in the use of prune trees for biological control protection in vineyards.

It can be useful to borrow concepts from landscape ecology in order to find more effective ways of providing habitat for beneficials and for managing agricultural pests. The study described herein explores the importance of changing the spatial structure of a vineyard landscape, particularly by establishing a vegetational corridor to enhance movement of beneficials beyond the "normal area of influence" of adjacent habitats or refuges. Corridors have long been used by conservation biologists for protecting biological diversity because they provide multiple avenues for circulation and dispersal of biodiversity through the environment (Rosenberg et al., 1997).

In northern California's Mendocino County, many vineyards are interwoven in a matrix of riparian forests. They provide many opportunities to study arthropod colonization and interhabitat exchange of arthropods, especially those restricted to the interstices between agricultural and uncultivated land.

This study took advantage of an existing 600-meter vegetational corridor composed of 65 flowering species. The corridor, which was connected to a riparian forest cutting across a monoculture vineyard, allowed for testing whether such a strip of vegetation could enhance the biological control of insect pests in a vineyard. We wanted to evaluate whether the corridor acted as a consistent, abundant, and well-dispersed source of alternative food and habitat for a diverse community of generalist predators and parasitoids, allowing predator and parasitoid populations to develop in the area of influence of the corridor well in advance of vineyard pest populations. We thought that the corridor would serve as a biological highway for the dispersion of predators and parasitoids within the vineyard, thus providing protection against insect pests.

Since the vineyard was diversified with cover crops, we could test another hypothesis: the presence of neutral insects, pollen, and nectar in summer cover crops provides a constant and abundant supply of food sources for natural enemies. This decouples predators and parasitoids from a strict dependence on grape herbivores, and allows natural enemies to build up in the system and keep pest populations at acceptable levels. We tested this hypothesis and examined the ecological mechanisms associated with insect pest reduction when summer cover crops were planted early in the season between alternate vine rows.

## 3.2 STUDY SITE

This study was conducted in two adjacent organic Chardonnay vineyard blocks (blocks A and B, 2.5 hectares each) from April to September in 1996 and 1997. Both vineyard blocks were surrounded on the north side by riparian forest vegetation.

Block A was penetrated and dissected by a 5-meter-wide by 300-meter-long vegetational corridor composed of 65 different species of flowering plants. The vineyard was located in Hopland, 200 kilometers north of San Francisco, California, in a typical wine-growing region. Before and during the study, both blocks were under organic management, planted yearly with winter cover crops every other row, receiving an average of 2 tons of compost per hectare and preventive applications of sulfur against *Botrytis* spp. and *Oidium* spp.

## 3.3 METHODS

### 3.3.1 Corridor

To determine whether the corridor influenced the species diversity and abundance of entomophagous insects in the adjacent vineyard, Malaise traps were placed across "flight paths" between block A and the corridor on the south side and the vineyard and the riparian forest on the north side. One malaise trap was also placed between block B of the vineyard and the adjacent bare edge. To maximize catches of flying and wind-carried arthropods at the vineyard interfaces, samples were taken from May through September. Each malaise trap contained a one quart glass jar filled with ethyl alcohol that was replaced every two weeks; it was taken into the laboratory for counting and sorting into families and trophic guilds.

Ten yellow and ten blue sticky traps were placed at different points (rows 1, 5, 15, 25, 45) within the vineyard at increasing distances from the corridor (block A) or the bare edge (block B) to monitor diversity and abundance of the entomofauna. Yellow sticky traps were used to monitor leafhoppers, the egg parasitoid *Anagrus epos*, and various predator species. Blue sticky traps were mainly used to assess thrips and *Orius* populations. Traps were oriented perpendicular to the predominant wind direction and positioned above the vine canopy. Traps were deployed beginning in April and replaced weekly throughout the 1996 and 1997 growing seasons. All traps were returned to the laboratory and examined with a dissection microscope to count the number of phytophagous insects and associated natural enemies in the traps.

Grape leaves were examined in the field and the number of *E. elegantula* nymphs recorded in the same rows where sticky traps were placed. Populations of leafhopper nymphs were estimated weekly on 10 randomly selected leaves in each row.

### 3.3.2 Cover Crop Blocks

Half of each block was kept free of ground vegetation by one spring and one late summer disking (the monoculture vineyard). In April, the other two halves of both blocks (the cover-cropped vineyard) were undersown every alternate row with a 30/70 mixture of sunflower and buckwheat. Buckwheat flowered from late May to July and sunflower bloomed from July to the end of the season.

From April to September of 1996 and 1997, relative seasonal abundance and diversity of phytophagous insects and associated natural enemies were monitored on the vines in both treatment plots. Ten yellow and ten blue sticky traps coated with tanglefoot (10 × 17 centimeters, Seabright Laboratories, Emeryville, CA) were placed in each of 10 rows selected at random in each block to estimate densities of adult leafhopper, thrips, *Anagrus* wasps, *Orius* spp., and other predators.

Grape leaves were examined in the field in the same rows where sticky traps were placed and the number of *E. elegantula* nymphs recorded. Populations of leafhopper nymphs were estimated on 10 randomly selected leaves in each row. This sampling method was carried out in sections with and without cover crops, allowing one to determine quickly and reliably the proportion of infested leaves, densities of nymphs, and rates of leafhopper egg parasitization by the *Anagrus* wasp (Flaherty et al., 1992).

Egg parasitism in vineyards was determined by examining the same 10 grape leaves under a dissection microscope for the presence of parasitized or healthy *E. elegantula* eggs. Unhatched eggs were examined for the presence of developing *A. epos* or *E. elegantula* (Settle and Wilson, 1990). At the same time, hatched leafhopper eggs were examined to determine presence of egg scars with round exit holes, indicating *A. epos* emergence (Murphy et al., 1996).

In order to determine whether cover crop mowing forced movement of natural enemies from cover crops to vines, three different selected cover crop rows in block B were subjected to mowing three times each year. Both years, five yellow and five blue sticky traps were placed in the three random rows with cover crops every time they were mowed, and in three random rows that were not mowed.

## 3.4 RESULTS

### 3.4.1 Influence of the Corridor on Leafhoppers and Thrips

In both years in block A, adult leafhoppers exhibited a clear density gradient, reaching lowest numbers in vine rows near the corridor and forest and increasing in numbers towards the center of the field, away from the adjacent vegetation. The highest concentration of leafhoppers occurred after the first 20 to 25 rows (30 to 40 meters) downwind from the corridor. Such a gradient was not apparent in block B, where the lack of the corridor resulted in a uniform dispersal pattern of leafhoppers (Figure 3.1); similar trends were observed in 1997. Nymphal populations behaved similarly, reaching highest numbers in the center rows of block A in both years. Apparently the area of influence of the corridor extended 15 to 20 rows (25 to 30 meters), whereas the area of influence of the forest on nymphs reached 10 to 15 rows (20 to 25 meters) as evident from 1997 catches. Nymphs were similarly distributed over the whole block-B field.

A similar population and distribution gradient was apparent for thrips (Figure 3.2); similar trends were observed in 1996. In both years catches in block A were substantially higher in the central rows than in rows adjacent to the forest; catches were particularly low in rows near the corridor. In block B there were no differences in catches between the central and bare edge rows, although catches near the forest were the lowest, especially during 1997.

Block A 1996

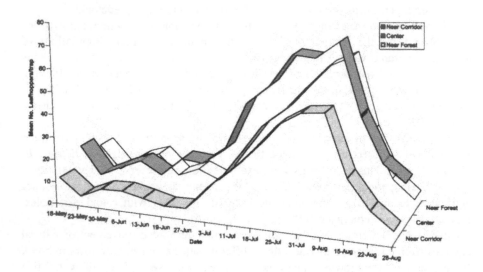

Block B 1996

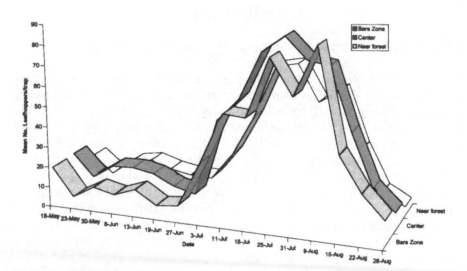

**Figure 3.1** Seasonal patterns (numbers per yellow sticky trap) of adult leafhopper *E. elegantula* in both vineyard blocks, as influenced by proximity to forest or the corridor (Hopland, California, 1996).

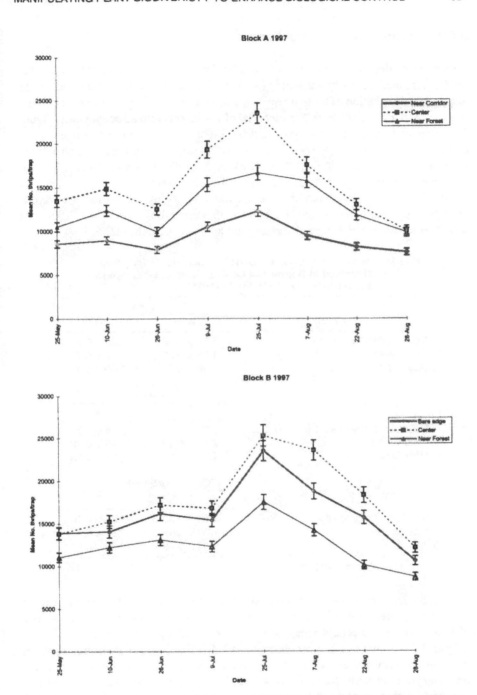

**Figure 3.2** Seasonal patterns of thrips (numbers per blue sticky trap) in both vineyard blocks, as influenced by proximity to forest or the corridor (Hopland, California, 1997).

### 3.4.2  Responses of Natural Enemies

Generalist predators in the families Coccinellidae, Chrysopidae, Nabidae, and Syr-phidae exhibited a density gradient in block A, clearly indicating that the abundance and spatial distribution of these insects were influenced by the presence of the forest and the corridor, which channeled dispersal of the insects into adjacent vines (Figure 3.3); similar trends were observed in 1996. Predators were more homogeneously distributed in block B; no differences in spatial pattern in predator catches were observed between bare edge and central rows, although their abundance tended to be higher in rows close to the forest (within 10–15 meters).

In block A the distribution of *Orius* sp. was affected by the corridor and forest, as higher numbers of *Orius* could be found in vines near the borders (up to 20 meters away), whereas in block B no dispersal gradient was apparent (Table 3.1).

Table 3.1   Mean (± SE) *Orius* sp. Densities per Blue Sticky Trap[a]
Observed in Border and Central Rows of Both Vineyard
Blocks in Hopland, California (1996)

| Location | June | |
|---|---|---|
| | A | B |
| Near corridor/bare edge | 1.33 ± 0.08 | 1.20 ± 0.3 |
| Field center | 1.16 ± 0.05 | 1.36 ± 0.45 |
| Near forest | 1.90 ± 0.47 | 1.40 ± 0.46 |
| | July | |
| | A | B |
| Near corridor/bare edge | 3.75 ± 0.94 | 2.54 ± 0.84 |
| Field center | 2.11 ± 0.52 | 2.96 ± 0.98 |
| Near forest | 4.52 ± 1.5 | 3.01 ± 0.75 |
| | August | |
| | A | B |
| Near corridor/bare edge | 1.53 ± 0.51 | 1.85 ± 0.56 |
| Field center | 1.20 ± 0.4 | 1.70 ± 0.62 |
| Near forest | 1.42 ± 0.38 | 2.03 ± 0.84 |

[a] Average of 4 sampling dates.

As *Anagrus* colonized grape vineyards from the corridor and forest throughout the sampling area, it exhibited higher densities in late July and throughout August of both years in the central vineyard rows where leafhoppers were most abundant (Figure 3.4); similar trends were observed in 1997. The increase in *A. epos* captures over time noticeable from late June onward indicated that parasitoids began moving into vineyards in early June, a few weeks after *E. elegantula* adults moved into vineyards. The appearance of *A. epos* coincided with the beginning of the oviposition period of leafhopper adults.

Leaf examination revealed high levels of parasitism across leafhopper generations for 1996 and 1997 in both blocks (Table 3.2). Eggs in center rows had slightly higher mean parasitization rates than eggs located in rows near the forest or corridor. The

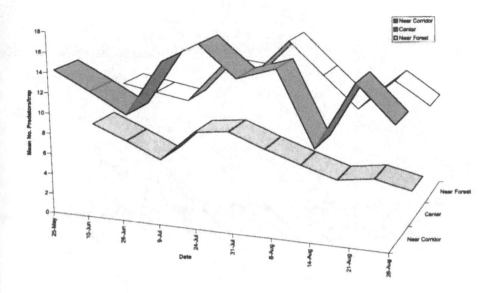

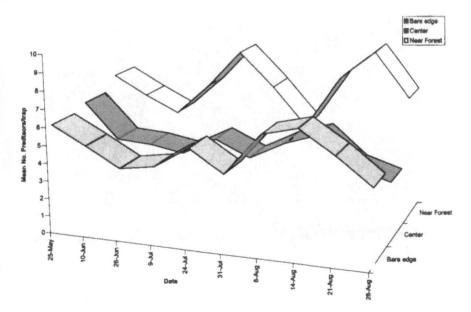

**Figure 3.3**  Seasonal patterns of predator catches (numbers per yellow sticky trap) in both vineyard blocks, as influenced by the proximity to forest or the corridor (Hopland, California, 1997).

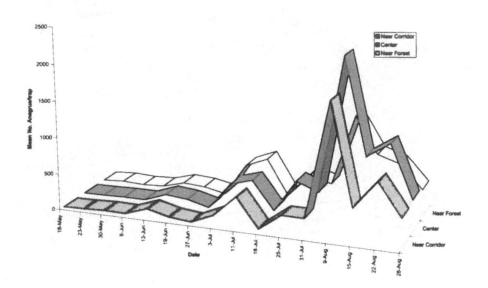

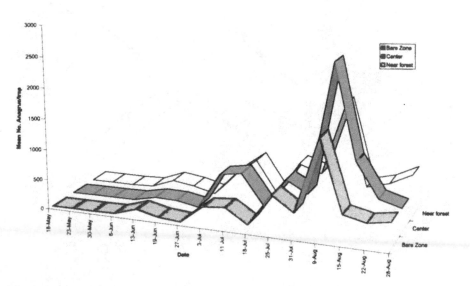

**Figure 3.4** Seasonal pattern of *Anagrus* catches (numbers per yellow sticky traps) in both vineyard blocks, as influenced by the proximity to forest or the corridor (Hopland, California, 1996).

Table 3.2    Mean (± SE) Percent Leafhopper Egg Parasitism[a] by *Anagrus epos* in
Border and Central Rows of Both Vineyard Blocks in Hopland, California

| Location | Block A | | Block B | |
|---|---|---|---|---|
| | 1996 | 1997 | 1996 | 1997 |
| Near corridor/bare edge | 46 ± 16 | 59 ± 14 | 62 ± 21 | 73 ± 45 |
| Field center | 61 ± 23 | 82 ± 33 | 75 ± 32 | 80 ± 37 |
| Near forest | 57 ± 31 | 77 ± 27 | 74 ± 43 | 75 ± 29 |

[a] Average of 12 sampling dates over the season.

proportion of eggs parasitized tended to be uniformly distributed across all rows in both blocks. It is assumed that the presence of the forest and corridor was associated with the colonization of *A. epos* but this did not result in a net season-long prevalence in *E. elegantula* egg parasitism rates in rows adjacent to such habitats.

### 3.4.3    Density Responses of the Grape Leafhopper to Summer Cover Crops

In both years, densities of adult leafhoppers were significantly lower throughout the season (except on June 27 and July 18 in 1996 and early in the summer in 1997) on vines with summer cover crops than on monoculture vines (Figure 3.5; $t = 2.612$, $df = 10$, $p < 0.05$).

Comparing the cover cropped vineyard with the monoculture shows that increasing plant diversity results in a decrease in the number of leafhopper nymphs. During 1996, nymphal densities were generally lower on vines in cover cropped sections. Differences were not statistically significant, however, from August 15 until the end of the season (Figure 3.6; $t = 2.31$, $df = 13$, $p < 0.05$). In 1997, significantly lower abundance levels of nymphs on cover cropped vines were evident from July 9 onward (Figure 3.6; $t = 2.50$, $df = 6$, $p < 0.05$).

### 3.4.4    Effects of Cover Crops on *Anagrus* Populations and Parasitization Rates

During 1996 the mean densities of *Anagrus* present on yellow sticky traps placed in cover cropped and monoculture vineyard sections were similar, although toward the end of the season *Anagrus* attained significantly greater numbers in the monoculture. Similarly during 1997, a year in which elevated capture rates were evident, sampling revealed significantly higher numbers of *Anagrus* in the monoculture starting in late July (Figure 3.7; $t = 2.41$, $df = 9$, $p < 0.05$). Clearly, *A. epos* was more abundant in the vineyard monocultures associated with higher host densities.

Differences in yellow sticky trap captures of *Anagrus* between cover-cropped and monoculture vineyards were not reflected in the parasitism records of *E. elegantula*. There was not a consistent relationship between leafhopper abundance and the measures of parasitism made in this study. No statistical differences in parasitization rates were detected between treatments in both years, although in

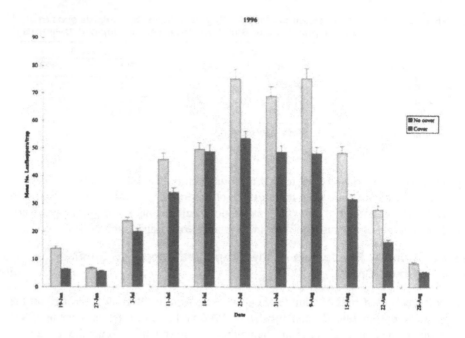

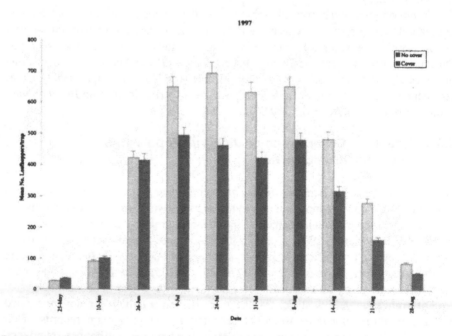

**Figure 3.5**  Densities of adult leafhoppers *E. elegantula* in cover cropped and monoculture vineyards in Hopland, California, during two growing seasons. Mean densities (number of adults per yellow sticky trap) and standard errors of two replicate means are indicated. In some cases error bars were too small to appear in the figure.

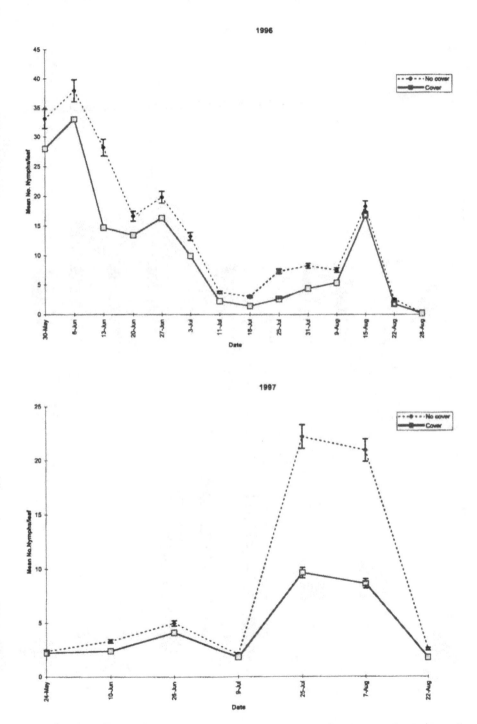

**Figure 3.6**  Densities of *E. elegantula* nymphs in cover cropped and monoculture vineyards during two growing seasons in Hopland, California.

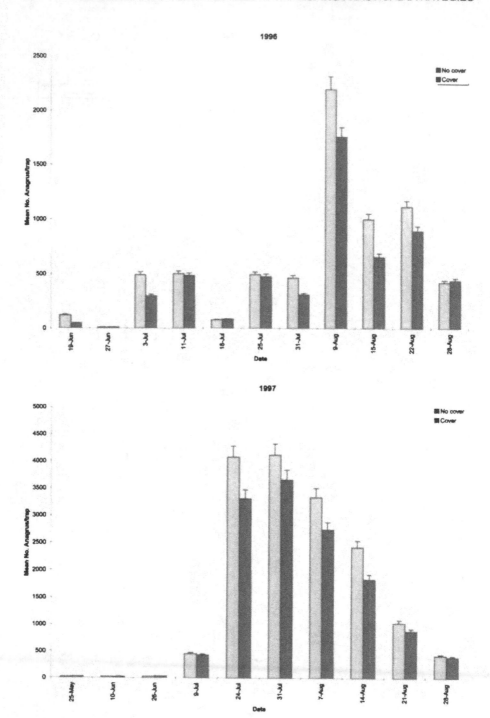

**Figure 3.7** Mean number of *Anagrus epos* per yellow sticky trap in cover cropped and monoculture vineyards during two growing seasons in Hopland, California.

Table 3.3   Mean Percent[a] Leafhopper Egg Parasitism by *Anagrus epos* during
Two Growing Seasons, in Vineyards with and without Cover Crops,
Hopland, California

| Month | | Cover cropped | No cover crop |
|---|---|---|---|
| 1996 | June | 48 a | 49 a |
| | July | 62 a | 59 a |
| | August | 67 a | 66 a |
| 1997 | June | 52 a | 54 a |
| | July | 64 a | 55 b |
| | August | 69 a | 68 a |

[a] Means in the same row followed by the same letter are not significantly different
(p <0.05, "t" test).

July of both years; egg parasitization was slightly higher, but not significantly so,
in the cover-cropped vineyard (Table 3.3; t = 3.67, df = 2, p <0.05).

## 3.4.5   Effects of Cover Crops on Thrips and General Predators

Densities of thrips, as revealed by blue sticky trap captures in 1996, were significantly lower (t = 2.37, df = 9, p <0.05) in cover cropped vineyards than in
monocultures, and remained lower throughout the growing season (Figure 3.8).
Such differences were also apparent in 1997, a year of extreme thrip pressure, as
thrip numbers were significantly greater in the monoculture starting in late July.
From this analysis, it is clear that an increase in vineyard plant diversity was
associated with lower thrip populations.

Table 3.4 shows the numbers of predators from cover cropped and monoculture
systems. The predators include spiders, *Nabis* spp., *Orius* spp., *Geocoris* spp.,
Coccinellidae, and *Chrysoperla* spp. Generally, the populations were low and
increased as prey became more numerous during the season. The table shows that
during 1996 general predator populations on the vines tended to be higher in the
cover cropped sections than in the monocultures.

D-Vac sampling of cover crops in both blocks revealed that in 1996 the most
abundant predator present on the flowers of buckwheat and sunflowers was *Orius*,
followed by several species of Coccinellidae. Among the spiders, members of the
family Thomisidae were the most common (Table 3.5). In 1997, *Orius* was again
the most abundant predator in the cover cropped sections, followed by several
thomisid spiders and a few species of Coccinellidae, Nabidae, and *Geocoris* spp.
Most of these predators probably responded to a complex of neutral insects, pollen,
and nectar present in the cover vegetation.

## 3.4.6   Effects of Cover Crop Mowing on Leafhoppers and *A. epos*

To determine whether mowing influenced leafhopper abundance in 1997, leafhopper
densities were assessed on vines selected before and after mowing; these densities were
then compared to leafhopper numbers on vines where cover crops were not mowed.

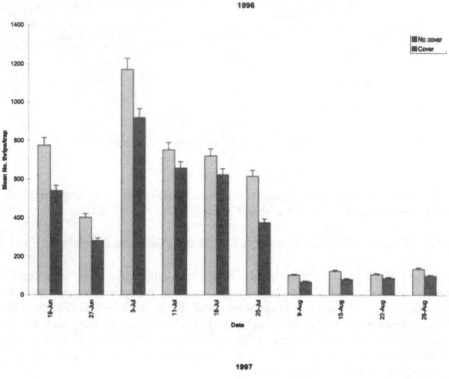

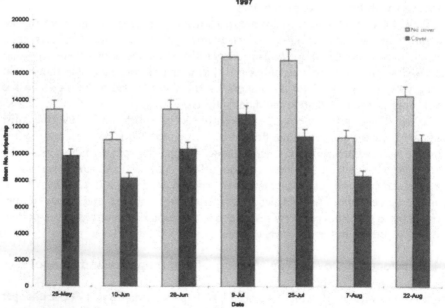

**Figure 3.8**   Mean densities of thrips per blue sticky trap in cover cropped and monoculture vineyards throughout two seasons in Hopland, California.

Table 3.4 Monthly Mean Densities[a] (± SE) of Various Arthropod Predator Species on Vines With and Without Summer Cover Crops, Hopland, California. 1996

| Cover Crop | Month | Orius | Spiders | Coccinellidae | Geocoris sp. | Nabis sp. | Chrysoperla sp. |
|---|---|---|---|---|---|---|---|
| With cover crop | June | 3 ± 0.7 | 3 ± 1.3 | 0 | 0 | 1 ± 0.3 | 3 ± 2.2 |
| | July | 5 ± 1.9 | 9 ± 3.4 | 4 ± 1.9 | 2 ± 1.7 | 1 ± 0.6 | 5 ± 3.1 |
| | Aug | 4 ± 2.0 | 12 ± 3.7 | 1 ± 0.8 | 4 ± 2.3 | 2 ± 1.1 | 2 ± 1.0 |
| Without cover crop | June | 2 ± 1.3 | 2 ± 1.1 | 2 ± 0.7 | 0 | 0 | 2 ± 0.7 |
| | July | 3 ± 0.9 | 8 ± 2.6 | 2 ± 0.4 | 1 ± 0.5 | 0 | 4 ± 1.5 |
| | Aug | 2 ± 0.8 | 9 ± 3.4 | 1 ± 0.3 | 2 ± 0.9 | 1 ± 0.7 | 2 ± 0.8 |

[a] Number of individuals per 25-m D-Vac transect.

Table 3.5 Proportion of Predator Groups Harbored by Summer Cover Crops (1996–1997) of Both Vineyard Blocks in Hopland, California (Caught by Sweep Netting, Average of 12 Sampling Dates over the Season).

| Year | Block A | | Block B | |
|---|---|---|---|---|
| 1996[a] | Orius sp. | 76% | Orius sp. | 68% |
| | Coccinellidae | 15% | Coccinellidae | 24% |
| | Others | 9% | Others | 8% |
| 1997[b] | Orius sp. | 83% | Orius sp. | 72% |
| | Spiders | 12% | Spiders | 17% |
| | Others | 5% | Others | 11% |

[a] Others in 1996 include Nabis sp., Geocoris sp., Chrysoperla sp. and several spider species.
[b] Others in 1997 include Coccinellidae, Nabis sp., Geocoris sp., and Chrysoperla sp.

Before mowing, leafhopper nymphal densities on vines were similar in the selected cover cropped rows. One week after mowing, numbers of nymphs declined on vines where the cover crop was mowed, coinciding with an increase in *Anagrus* densities in mowed cover crop rows. During the second week, this decline was even more pronounced (t = 2.93, df = 4, p <0.05), although by then differences in *Anagrus* numbers between mowed and not mowed rows were not significant (Figure 3.9).

## 3.5 CONCLUSIONS

Our studies showed that cover crops harbored large numbers of *Orius*, coccinellids, thomisid spiders, and a few other predator species. Comparisons of predator abundance in both blocks showed that the presence of the buckwheat and sunflower produced an increase in the density of predators. This result is consistent with the observations reported by Daane and Costello (1998), who found that cover crops influenced the relative abundance of spiders present in vineyards. The question is whether such enhancements in predator abundance (especially because *Anagrus* behaved similarly in both systems) explains the lower populations of leafhoppers and thrips detected in the diversified vineyards. Some researchers (Hanna et al., 1996) believe that leafhopper reductions may be attributed to enhanced activity of a certain group of spiders, which are consistently found at higher densities in the presence of cover crops compared to clean cultivated systems. Our analysis reveals that greater densities of predators are correlated with lower leafhopper numbers; this relationship is most obvious in the *Orius*–thrips interaction.

The mowing experiment suggests a direct ecological linkage. The cutting of the cover crop vegetation forced the movement of the *Anagrus* and predators harbored by the flowers, which resulted in a decline of leafhopper numbers on the vines adjacent to the mowed cover crops in both years. These results are consistent with findings of Sluss (1967), who recommended cutting cover crops in walnut orchards in late April or early May to force movement of *Hippodamia convergens* onto the walnut trees to exert early control of the walnut aphid. Clearly, more research is needed on the timing of mowing in relation to the biology of the leafhopper and the phenology of the vine and cover crops.

This research indicates that dispersal and subsequent within vineyard densities of herbivores and associated natural enemies are influenced by adjacent landscape features such as forest edges and corridors. The presence of riparian habitats enhances predator colonization of and abundance in adjacent vineyards, although this influence is limited by the distance to which natural enemies can disperse into the vineyard (Corbett and Plant, 1993). The corridor, however, amplifies this influence by allowing enhanced and timely circulation and dispersal movement of predators into the center of the field. The great availability of pollen and nectar displayed by the various flowers of the corridor, as well as the diversity and prevalence of neutral insects, attracted high numbers of generalist predators. Increased abundance of alternative food has often been associated with a rise in predator abundance, apparently enhancing predators' reproduction and/or survival (Lys et al., 1994). This increased predator abundance in turn increases the impact

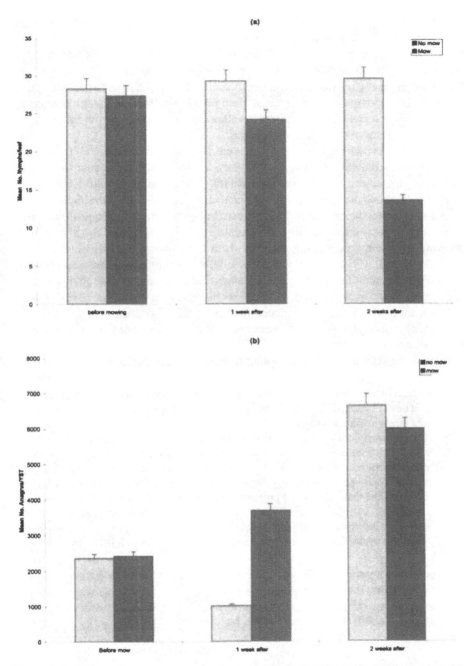

**Figure 3.9**  Effects of cover crop mowing in vineyards on densities of (a) leafhopper nymphs and (b) *Anagrus epos* during the 1997 growing season in Hopland, California.

of predators, especially in crop rows close to habitats providing alternative food (Coombes and Sotherton, 1986).

Many of the predator species present in the corridor originated from the riparian forest edge. For some predators, such as Coccinellidae, Chysopidae, and Syrphidae, the corridor influenced numbers and dispersal in late spring and early summer, the effect acting through the presence of noncrop aphids and other Homoptera (for Coccinellidae and Chrysopidae) and nectar and pollen (for Syrphidae). Some plant species harbored populations of neutral Homoptera and Hemiptera, which acted as important food reservoirs for predatory Anthocoridae and Miridae migrating from the forest and later moving into the vineyard.

Unlike the predator species, the parasitoid *A. epos* was not affected by vegetational diversity directly. Although it has been shown that *A. epos* colonizes vines from vineyard edges (Corbett and Rosenheim, 1996), the parasitoid in this study followed the abundance patterns of leafhoppers and did not display the distributional response exhibited by predators. Other researchers who have found a positive effect of flowers on parasitoid diversity and abundance have also reported the difficulty of showing an evident gradient of parasitoids from a rich flowering habitat into a crop area (Duelli et al., 1990). Given that *A. epos* dispersed similarly across rows in both blocks, it is apparent that predator enhancement near the vegetational interfaces best explains the lower populations of leafhoppers and thrips in the border rows of block A. Such successful impact of predators can be assumed because fewer adults and nymphs of leafhoppers and thrips were caught near the corridor than in the middle of the vineyards.

The data obtained in this study point to two main conclusions:

- Habitat diversification using summer cover crops supports season long high populations of predators, thereby favoring enhanced biological control of leafhoppers and thrips in vineyards.
- The creation of corridors across vineyards can serve as a key strategy for allowing natural enemies emerging from riparian forests to disperse over large areas of otherwise monoculture systems. Such corridors should be composed of locally adapted plant species exhibiting sequential flowering periods, which attract and harbor an abundant diversity of predators and parasitoids. These corridors or strips, which may link various crop fields and riparian forest remnants, can create a network of habitat allowing many species of beneficial insects to disperse throughout whole agricultural regions, transcending farm boundaries (Baudry, 1984).

Our study suggests that it is possible to restore natural controls in agroecosystems through vegetation diversification, thus providing a robust ecological foundation for the design of pest stable and sustainable vineyards in northern California and in the Mediterranean world.

## REFERENCES

Altieri, M.A., *Biodiversity and Pest Management in Agroecosystems*, Haworth Press, New York, 1994.

Andow, D.A., Vegetational diversity and arthropod population response, *Ann. Rev. Entomol.* 36, 561–586, 1991.

Baudry, J., Effects of landscape structure on biological communities: the cases of hedgerows network landscapes, in Brandt, J. and Agger, P., Eds., *Methodology Landscape Ecological Res. Plan.*, Vol. 1, Roskilde University Center, Denmark, 1984, 55–65.

Boller, E.F., The ecosystem approach to plan and implement integrated plant protection in viticulture of eastern Switzerland, in Cavalloro, R., Ed., Plan-Protection Problems and Prospects of Integrated Control in Viticulture, *Proceedings of the International CEC-IOBC Symposium*, Lisboa, Portugal, Report EVR 11548, 1990, 607–617.

Bugg, R.L and Waddington, C., Using cover crops to manage arthropod pests of orchards: a review, *Agricul. Ecosystems Environ.*, 50, 11–28, 1994.

Corbett, A. and Rosenheim, J.A., Impact of natural enemy overwintering refuge and its interaction with the surrounding landscape, *Ecol. Entomol.*, 21, 155–164, 1996.

Corbett, A. and Plant, R.E., Role of movement in the response of natural enemies to agroecosystem diversification: a theoretical evaluation, *Environ. Entomol.*, 22, 519–531, 1993.

Coombes, D. S. and Sotherton, N.W., The dispersal and distribution of polyphagous predatory Coleoptera in cereals, *Ann. Appl. Biol.*, 108, 461–474, 1986.

Daane, K.M., Costello, M.J., Yokota, G.Y., and Bentley, W.J., Can we manipulate leafhopper densities with management practices? *Grape Grower*, 30(4), 18–36, 1998.

Daane, K.M. and Costello, M.J., Can cover crops reduce leafhopper abundance in vineyards? *Calif. Agric.*, 52(5), 27–32, 1998.

Doutt, R. and Nakata, J., The *Rubus* leafhopper and its egg parasitoid: an endemic biotic system useful in grape-pest management, *Environ. Entomol.*, 2, 381–386, 1973.

Duelli, P., Studer, M., Marchand, I., and Jakob, S., Population movements of arthropods between natural and cultivated areas, *Biol. Conserv.*, 54, 193–207, 1990.

Flaherty, D.L., Ecosystem trophic complexity and the Willamette mite, *Eotetranychus willamettei* (Acarine: Tetranychidae) densities, *Ecology*, 50, 911–916, 1969.

Flaherty, D.L., Christensen, P.T., Lanini, T., Marois, J., and Wilson, L.T., *Grape Pest Manage.*, University of California Division of Agriculture and Natural Resources, 1992.

Fry, G., Landscape ecology of insect movement in arable ecosystems, in Glen, D.M., Greaves, M.P., and Anderson, H.M., Eds., *Ecol. Integrated Farming Syst.*, John Wiley & Sons: Bristol, UK, 177–202., 1995.

Hanna R., Zalom, F.G., and Elmore, C.L., Integrating cover crops into vineyards, *Grape Grower*, 26–43, February 1996.

Kido, H., Flaherty, D.L., Bosch, D.F., and Vaero, K.A., French prune trees as overwintering sites for the grape leafhopper egg parasite, *Am. J. Enol. Vitic.*, 35, 156–160, 1984.

Lewis, T., The effects of shelter on the distribution of insect pests, *Scientific Hortic.*, 17, 74–84, 1965.

Lys, J.A., Zimmermann, M., and Nentwing, W., Increase in activity density and species number of carabid beetles in cereals as a result of strip-management, *Entomol. Exp. Appl.*, 73, 1–9, 1994.

Murphy, B.C., Rosenheim, J.A., and Granett, J., Habitat diversification for improving biological control: Abundance of *Anagrus epos* (Hymenoptera: Mymaridae) in grape vineyards, *Environ. Entomol.*, 25(2), 495–504, 1996.

Pollard, E., Hedges IV. A comparison between the carabidae of a hedge and field site and those of a woodland glade, *J. Appl. Ecol.*, 5, 649–657, 1968.

Rosenberg, D.K., Noon, B.R., and Meslow, E.C., Biological corridors: Form, function and efficacy, *BioScience*, 47(10), 677–687, 1997.

Settle, W.H. and Wilson, T., Invasion by the variegated leafhopper and biotic interactions: Parasitism, competition, and apparent competition, *Ecology*, 71, 1461–1470, 1990.

Settle, W.H., Wilson, L.T., Flaherty, D.L., and English-Loeb, M., The variegated leafhopper, an increasing pest of grapes, *Calif. Agric.*, 40, 30–32, 1986.

Sluss, R.R., Population dynamics of the walnut aphid *Chromaphis juglandicola* (Kalt) in northern California, *Ecology,* 48, 41–58, 1967.

Sotherton, N.W., The distribution and abundance of predatory arthropods overwintering on farmland, *Ann. Appl. Biol.,* 105, 423–429, 1984.

Thomas, M.B., Wratten, S.D., and Sotherton, N.W., Creation of "islands" habitats in farmland to manipulate populations of biological arthropods: Predator densities and emigration, *J. Appl. Ecol.,* 28, 906–917, 1991.

Wratten, S.D., The role of field margins as reservoirs of natural enemies, in Burn, A.J., Ed., *Environmental Management in Agriculture,* Belhaven Press, London, 1988.

CHAPTER **4**

# An Assessment of Tropical Homegardens as Examples of Sustainable Local Agroforestry Systems

V. Ernesto Méndez

## CONTENTS

1-8493-0894-1/01/$0.00+$.50
© 2001 by CRC Press LLC

## 4.1 INTRODUCTION

The need to include local knowledge in sustainable agricultural research and development has been increasingly recognized (Altieri, 1995; Madge, 1995; Sinclair and Walker, 1999). This has been particularly important in the tropics, where experience has demonstrated the shortcomings of introducing agricultural technologies that are not ecologically or culturally adapted to the local environment (NRC, 1993). In response to these negative results, many researchers in tropical areas have focused their attention on the potential of locally developed agroecosystems (Gliessman et al., 1981; Altieri and Anderson, 1986; Schulz et al., 1994). In addition to having positive agroecological qualities, locally developed agroecosystems can also provide insight into practices and extension strategies that are more acceptable and more easily adopted by farmers (Rocheleau, 1987).

There are two main reasons why locally developed agricultural systems are interesting subjects of study for sustainable agricultural research. First, many local agroecosystems have withstood the test of time, albeit with continual modifications and adaptations (Ellis and Wang, 1997). Second, these agricultural systems are usually well adapted to local ecological and social realities, and depend mostly on available renewable resources (Klee, 1980; Marten, 1986; Wilken, 1988).

Agroforestry systems* stand out as one of the most ancient and widespread practices in the tropical regions (Nair, 1989) in the diversity of locally developed and indigenous agroecosystems in the world today. Scientific understanding of the biophysical components and interactions present in tropical agroforestry systems has increased greatly in the last two decades (Sanchez, 1995). Unfortunately, insufficient attention has been placed on research that focuses on the socioeconomic and cultural contexts that affect the success of agroforestry as a land use option (Nair, 1993; Scherr, 1995; Nair, 1998). As Rocheleau (1999) has pointed out, it is necessary to evaluate equally the ecological, socioeconomic, and cultural characteristics of tropical agroforestry systems in order to determine their sustainability.

Rather than discuss locally developed agroforestry systems in general, this chapter will focus on a particular type of agroforestry system — the tropical homegarden. Found in many parts of the world, tropical agroforestry homegardens (referred to as homegardens) can be defined as land use systems that include deliberate associations of trees, herbaceous crops, and/or animals in close interaction with a household (Fernandes and Nair, 1986).

Homegardens have been described as containing characteristics of sustainable agricultural systems by numerous authors (Torquebiau, 1992; Jose and Shanmugaratnam, 1993; Gliessman, 1998). However, their ecological complexity and the strong interaction that exists between the agroecosystem and the household have made it difficult for researchers to conduct in depth studies that would make these claims conclusive (Wojtkowski, 1993).

---

* Agroforestry can be defined as a land use system combining trees with agricultural crops and/or animals, in which ecological interactions are managed in order to obtain multiple social, economic and/or environmental products and benefits (adapted from Nair, 1993 and Somarriba, 1998).

The first section of this chapter reviews selected studies that have attempted to analyze the sustainable qualities of tropical homegardens in different settings. The main objective of this section is to show the reader the nature of the information that is available on homegarden sustainability. The following section presents a case study that analyzes the interaction between some ecological and socioeconomic components of homegardens in Nicaragua. This section aims to present a detailed picture of the nature of tropical homegardens. It also demonstrates the value of the information that can be gained by using interdisciplinary research approaches. The final section discusses the research that will need to be done to better assess the sustainability of homegardens and similar agroforestry systems in the tropics.

## 4.2 TROPICAL HOMEGARDENS AND SUSTAINABILITY

Torquebiau (1992) has presented the most complete literature review on homegarden sustainability. This review draws on data from a variety of studies to test several descriptors of sustainability, each with a series of empirical indicators. The review concludes that homegardens contain the following broad attributes of sustainable agroecosystems:

1. Conservation of soil fertility and erosion control
2. Modification of the microclimate
3. Uniform and diversified production throughout the year
4. Use of endogenous inputs
5. Management flexibility
6. Diverse social roles
7. Limited impact on other systems.

Although the work of Torquebiau provides a valuable analysis that points to the characteristics of homegardens, it also shows the descriptive nature of most of the information that was available.

Other publications from the early 1990s contributed similar information. Landauer and Brazil (1990) published results from an international conference on tropical homegardens. This volume is an important source of information on homegardens from around the world. A chapter of great relevance to studies of sustainability is that by Michon and Mary (1990). This chapter describes changes in the structure and species composition of four homegardens in Java and Sumatra that occurred in response to local socioeconomic and demographic pressures. They report that the expansion of urban centers led to the opening of markets for products that were not traditionally grown in homegardens. In general, the number of plant species decreased (from 50 to 15), as did the number of vertical strata (from 4 to 5 to 2 to 3). Similar changes are becoming increasingly common in most parts of the tropical world where homegardens exist, a situation that merits closer attention (Hoogerbrugge and Fresco, 1993). The scenario presented here raises questions as to the long-term survival of homegardens when the families who maintain them are strongly influenced by external social and economic forces.

Gliessman (1990) attributed to Mexico and Costa Rica homegardens the following characteristics of sustainability: structural and productive diversity, maintenance of the resource base, and similarity to the local natural system. This work included information on a variety of the ecological characteristics of the homegardens that were studied, including size, plant diversity (categorized by use and habit), leaf area index, percent cover, and light transmission.

Jose and Shanmugatnam (1993) described the homegardens of Kerala as having chronological and structural characteristics similar to those of tropical forests. This study included an analysis of the vertical and horizontal structures of homegardens, including canopy cover, horizontal spatial arrangements, and species composition in each of the vertical strata. The chronological development was evaluated by determining the age of the trees inside the homegardens (which ranged from less than 5 years to 20 years). In addition, the study presents significant socioeconomic information on the perceived benefits and services farmers obtain from homegardens. The farmers interviewed in this study assigned significant importance to homegardens as sources of food and family habitat. According to the authors, the homegardens of Kerala are representations of low to medium input sustainable agroecosystems.

Jensen (1993a) provided a very complete analysis of homegarden soil conditions (bulk density, texture, conductivity, pH, CEC, organic matter, etc.) and nutrient concentrations in soil and biomass. The study took place in one Javanese homegarden over an 8 month period. Based on this analysis, the author stresses the importance of homegardens as low input agroecosystems that may stabilize sloping land and contribute to water and soil fertility conservation in Java. However, he also points to the need for more research in order to make more conclusive evaluations. In another paper, Jensen presents an analysis of productivity and nutrient cycling in the same homegarden (1993b). Here, the author draws more conclusive arguments as to the efficient cycling of nutrients observed in the homegarden, which allows for production without the use of external fertilizers and pesticides.

Lok (1998a) presented a collection of multidisciplinary studies of homegardens in Central America. The book is composed of several case studies that discuss many factors that are important for research on homegarden sustainability. For example, there are chapters on water conservation issues and management of animals in homegardens, two topics that have been largely ignored to date. The presentation of both socioeconomic and agroecological data in most of the cases also adds great value to this publication. Although sustainability is not addressed directly, the book contains valuable information on different ways of integrating and analyzing information on the ecological and socioeconomic characteristics of homegardens.

Gajaseni and Gajaseni (1999) have presented one of the most recent studies dealing with the ecological sustainability of homegardens. Their paper on homegardens in Thailand which have been continuously managed for at least three generations, contains several important contributions to sustainability research. The manuscript includes quantitative analysis of soil and vegetation data that has not been previously documented in the literature. Most of this information was compared to figures from local forest ecosystems, which were used as a baseline for sustainability within the local environment. The authors evaluated the homegardens based on five indicators of ecological sustainability:

1. Locally developed ecological knowledge base
2. Physical structure
3. Biological diversity
4. Nutrient cycling
5. Microenvironment as compared to the environment outside the homegarden (air and soil temperature, and relative humidity).

The homegardens showed favorable, if somewhat different, results for all indicators when compared to forest ecosystems. The use of a local ecological rationale for homegarden management demonstrated their evolution as part of the knowledge system present in the area. The main focus of this knowledge is to reproduce forest structure, while substituting forest species for those that are most useful to humans. In this respect the authors note a very different species composition between the forest and homegardens, while the physical structure of the two were observed to be very similar. The preservation of the forest's structure resulted in efficient nutrient cycling, which allowed for the maintenance of soil fertility without the use of synthetic fertilizers. In addition, no threatening pest outbreaks were reported.

The studies presented above illustrate the nature of most of the research on homegarden sustainability carried out in the last decade. All of these investigations contributed valuable information on homegarden characteristics and on their potential as sustainable agroecosystems. However, few publications were able to present comparative, quantitative data over medium- to long-term periods. In part, this is due to the difficulty of obtaining quantitative measurements from a system with such a high level of structural and temporal complexity (Mergen, 1987; Wojtkowski, 1993). On the other hand, it shows the lack of institutional and financial support required to carry out more in depth studies of agroecosystems with these characteristics (Nair, 1993). Nevertheless, an evolution towards more in depth and interdisciplinary analyses can be traced chronologically in the publications discussed above. Future studies can begin to provide the necessary information to better demonstrate the sustainability (or unsustainability) of these complex systems.


## 4.3 HOMEGARDENS IN NICARAGUA: AN INTERDISCIPLINARY CASE STUDY

This section summarizes the results of a case study examining the relationship between the social and economic importance of homegardens and their agroecological characteristics (Méndez, 1996; Méndez, Lok and Somarriba, 1999; Méndez, Lok, and Somarriba, 2000). The study's main objective was to gain an understanding of the rationale behind the design and management of homegardens. This type of research is essential for studies of sustainability because it provides insight into what causes farmers to maintain or discard management practices that have an impact on the sustainability of their agroecosystems.

### 4.3.1    Description of the Study Site

Research was conducted in the town of San Juan de Oriente, located in the semidry tropical zone of Nicaragua at an altitude of 450 m. Mean annual precipitation and temperature are 1500 mm and 26°C, respectively. The main agricultural products at the time of the study were coffee (*Coffea* spp.), ornamental plants, and fruit trees. Several types of bananas and plantains (*Musa* spp.), maize (*Zea mays*), and beans (*Phaseolus vulgaris*) are also grown for local consumption. The town's proximity to Managua, Nicaragua's capital, and two other important urban centers (Granada and Masaya), ensured adequate markets for agricultural products and handicrafts (important economic activities at the village). High population density (358 inhabitants/km$^2$) has caused increased pressure on land and resources in the region (Lok, 1994). Twenty homegardens were selected for the study, with areas ranging between 0.02 ha and 2 ha, with an average of 0.23 ha. Most of the homegardens (17) had areas under 0.5 ha, with only three units encompassing larger areas (1 to 2 ha).

### 4.3.2    Methodology

Information on the homegardens was compiled from January to August 1996. Data on zonation (allocation of certain areas to specific uses) and plant use were collected through participatory mapping and plant inventories. Zones were described by farmers according to their main functions, measured (in m$^2$), and expressed as percentage of total area. Most zones and plants had multiple functions and uses. For classification purposes, the primary zone functions and plant uses as reported by the farmers were used. Socioeconomic data was collected through surveys, direct observation, and semistructured interviews. A cluster analysis using Ward's minimum variance method (SAS Institute, 1987) was used to identify homegarden types, using the following variables: (1) number of zones, (2) number of plant uses, (3) number of plant species, and (4) total area. Groups determined by the cluster analysis were compared against types defined on a functional basis prior to the statistical analysis, and which matched the cluster analysis by 85%. This initial nonstatistical typology was based on data from the entire field period, including that collected during surveys, observations, and interviews. Case studies were done with three homegardens, each of a different type, as separated by the cluster procedure. The objective of the case studies was to collect more in depth information (mostly through semistructured and informal interviews) on homegarden and family history, as well as the caretaker's knowledge of homegarden management.

### 4.3.3    Results

#### 4.3.3.1    Management Zones

Ten management zones were identified in the 20 homegardens (Table 4.1). The most frequent zones were those identified as residential, fruit trees, ornamentals with

Table 4.1  Frequency and Average Percentage of Total Area of Homegarden
Management Zones

| Management Zone | Frequency (n = 20) | Zone as % of total area |
|---|---|---|
| 1. Fruit trees | 13 | 37 |
| 2. Shaded coffee | 6 | 16 |
| 3. Residential | 20 | 25 |
| 4. Ornamentals with shade trees | 12 | 14 |
| 5. Multi-purpose trees[a] | 5 | 3 |
| 6. Herbaceous crops[b] | 2 | 1 |
| 7. Ornamentals with vine-crop shade | 4 | 1 |
| 8. Grass[c] | 1 | 0.4 |
| 9. Other | 3 | 2 |
| 10. Ornamentals with artificial shade | 4 | 1 |

[a] Trees used for fuelwood, timber, forage, and as ornamentals.
[b] Herbaceous food crops and medicinal plants.
[c] Used during the firing of certain ceramics.

Source: From Méndez, V.E., Influéncia de factores socieconómicos sóbre estructuras agroecológicas des huertos caseros en Nicaragua, M.S. thesis, CATIE, Turrialba, Costa Rica, 1996. With permission.

shade trees, and shaded coffee. On average, the greatest proportion of homegarden area was allocated to fruit trees (37%) and residence (25%). Shaded coffee and ornamentals with shade trees were allocated 14 to 16% of the total area when they were present. Zone location was usually deliberate and based on practical considerations, plant requirements, and soil and microclimatic conditions. For example, ornamentals and herbaceous crops were close to the household to facilitate watering, weeding, safeguarding, and direct sales (Figure 4.1).

### 4.3.3.2  Plant Diversity and Use

A total of 324 plant species for nine different uses were identified (Table 4.2). Plant species diversity in each homegarden ranged between 22 and 106, with an average of 70. Fruit production, medicinal plants, Musa spp. fruit production, multipurpose trees, ornamental plants, and timber trees were the most frequent uses, each present in at least 85% of the sample. The ornamental use category contained the highest species diversity, followed by fruit trees and multipurpose trees. A total of 85 tree species were used as sources of fruit, timber, and other multiple uses (firewood, posts, medicinal, etc.).

### 4.3.3.3  Occupation by Gender

Nine occupations, four of which were divided equally between two activities, were reported by 85% of the sample population (Table 4.3). Homegarden management was the third most frequent occupation for both sexes. In the case of the men, it was somewhat more important, since an additional 22% reported homegarden management as a half time activity.

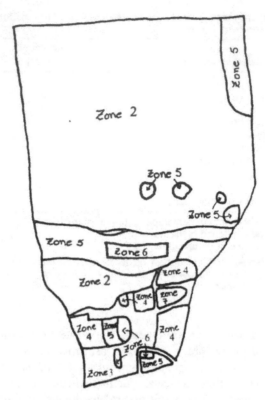

**Figure 4.1**   Map of homegarden showing six different management zones. Zone 2: shaded
coffee; Zone 3: residential; Zone 4: ornamentals with shade trees; Zone 5: multipurpose trees; Zone 6: herbaceous crops; Zone 7: ornamentals with vine crop
shade. From Méndez, Lok, and Somarriba (2000).

**Table 4.2   Plant Use, Frequency, and Species Diversity in 20 Nicaraguan
Homegardens**

| Plant use | Frequency | Total number of species |
|---|---|---|
| Fruit production (trees) | 20 | 37 |
| Multipurpose (trees) | 20 | 35 |
| *Musa* spp. fruit production | 20 | 3 |
| Ornamental (herbaceous plants)[a] | 19 | 180 |
| Timber/construction (trees)[b] | 19 | 14 |
| Medicinal (herbaceous plants) | 18 | 24 |
| Food (herbaceous crops) | 17 | 9 |
| Food/Spice (perennial shrubs) | 15 | 3 |
| Multipurpose (herbaceous plants) | 10 | 19 |
| Sample total | — | 324 |

[a] Includes annual and perennial herbaceous plants.
[b] Includes timber trees and bamboo.

Adapted from Méndez, V.E., Lok, R., and Somarriba, E., Interdisciplinary analysis
of homegardens in Nicaragua: zonation, plant use and socioeconomic importance, *Agroforestry Syst.* (in press).

Table 4.3  Occupations by Gender Reported by 20 Nicaraguan Households with Homegardens

| Occupation | % of Women | % of Men |
|---|---|---|
| Homegarden management | 16 | 10 |
| Student | 50 | 37 |
| Handicrafting | 21 | 8 |
| Outside work | 8 | 20 |
| Household work | 5 | 0 |
| Handcrafting and homegarden management | — | 5 |
| Outside work and homegarden management | — | 7 |
| Student and homegarden management | — | 10 |
| Student and outside work | — | 3 |

*Source*: From Méndez, V.E., Lok, R., and Somarriba, E., Interdisciplinary analysis of home-gardens: a case study from Nicaragua, in Jimenez, F. and Beer, J., (Eds.), *Proceedings of the International Symposium on Multi-Strata Agroforestry Systems with Perennial Crops*, CATIE, Turrialba, Costa Rica, 1999.

### 4.3.3.4  Use of External Inputs

Only two of the homegarden households reported the use of synthetic fertilizers or pesticides. These families utilized one fertilizer application a year of an N-P-K formula (at unknown concentrations) for coffee and passion fruit production. Most of the other families stated that they could not afford to buy these products.

### 4.3.3.5  Labor Investment

An average of three individuals per family regularly contributed labor to home-garden management, and these homegarden caretakers were distributed almost equally between men (52%) and women (48%). Average reported labor input was 32.6 hours per week. The amount of labor invested per family in homegarden management varied according to family size and occupation. Labor inputs by gender were variable, and seemed to depend more on the number of women and men than on assigned gender roles. In only one homegarden were tasks defined by gender. The men of the family were in charge of fruit trees, coffee, and other crops, while the women attended exclusively to the cultivation and sales of orna-mental plants.

### 4.3.3.6  Products and Benefits

A total of 40 plant products for consumption and sales were obtained from home-gardens. Most frequent were fruits, especially different types of oranges and lemons (*Citrus* spp.), mangoes of different varieties (*Mangifera indica*), avocados (*Persea americana*), and coconuts (*Cocos nucifera*). Other products that were frequently reported included *Musa* spp. for consumption, and coffee, passion fruit (*Passiflora* spp.), and ornamental plants for sale. Nine families cited space for handicrafting as an important use of the homegarden. All of the families acknowledged the importance of the homegarden as a place to work, relax, and socialize.

### 4.3.3.7   Income Generation

Four main sources of income were reported by the homegarden families: (1) home-garden products, (2) handicrafting, (3) commerce,* and (4) outside labor. The most important income source was homegarden products, which was mentioned by 70% of the sample (n = 20). Income from homegarden products averaged 34.5% of total income for all the families. The second and third most important income sources were handicrafting and outside work, cited by half of the families and representing an average of 32.5% and 26.5% of total income respectively.

### 4.3.3.8   Classification of Homegardens

The classification procedure identified six types of homegardens, using four agro-ecological variables (Table 4.4). All variables made a significant (p <0.001) con-tribution to the clustering procedure, but zonation (Z) and total area (A) had the most weight in the classification. In addition, important differences were found in the average percentage of total income and weekly labor inputs for each homega-rden type.

Ornamental homegardens (type A) were small and specialized in the production of ornamental plants for commerce. Most of their area was allocated to ornamental zones, and their plant diversity was concentrated in the ornamental category. Income generation from the homegarden was relatively high as compared to the other types, and labor inputs were in the medium range.

Handicrafting homegardens (type B) were small, had the fewest plant species, and together with type A homegardens, the fewest number of zones. Plant compo-nents in these homegardens were mixtures of different types of trees and bananas. Handicrafting was the principal activity and the homegardens were important in the space they provided for working and the shade they provided for drying the hand-icrafts. The lowest amount of labor investment was observed in this type of garden, and no direct income was reported from homegarden products.

Subsistence homegardens (type C) had areas in the medium range as compared to the other types. Homegarden products were used mostly in the households. There were relatively few management zones and medium plant species diversity compared to the rest of the sample. Few homegarden products were sold, and a very low percentage of income came from homegarden products. Labor investments were similar to those of the ornamental type.

Handicrafting and mixed production homegardens (type D) produced handicrafts and plants for sale and for household use. These homegardens were of medium size and included relatively high numbers of plant species and management zones. Labor inputs in this category were the highest for the entire sample. This could be attributed to the many individuals involved in different aspects of homegarden management. Income was generated from several sources. The portion coming from homegarden products was medium as compared to the other types.

---

* This category is composed of intermediaries in the sales of local products, mainly handicrafts and ornamental plants. The bulk of the products come from homegardens or households other than their own.

Table 4.4  Averages of Variables Used for Classification (S, U, Z, and A), and Average Income Generation Percentages and Labor Inputs for Each of the Types Defined by the Cluster Procedure

| Homegarden types | Total number of plant species (S) | Number of plant uses (U) | Number of management zones (Z) | Total area, m² (A) | Percent of income direct from homegarden | Weekly labor input (hrs) |
|---|---|---|---|---|---|---|
| A. Ornamental | 75 | 7 | 2 | 499 | 60 | 24 |
| B. Handcrafting | 35 | 7 | 2 | 672 | 0 | 9 |
| C. Subsistence | 48 | 8 | 3 | 1705 | 2 | 28 |
| D. Handicraft and mixed production | 89 | 9 | 4 | 1852 | 40 | 52 |
| E. Mixed production | 95 | 9 | 6 | 7838 | 75 | 41 |
| F. Minimal management | 96 | 9 | 3 | 14000 | 10 | 36 |

Adapted from Méndez, V.E., Lok, R., and Somarriba, E., Interdisciplinary analysis of homegardens in Nicaragua: zonation, plant use and socioeconomic importance, *Agroforestry Syst.* (in press).

Mixed production homegardens (type E) were large and contained the highest diversity of plant species and management zones. Labor inputs were high and income generated from the homegarden was the highest among all types. The most important zones, shaded coffee and ornamental plants, were maintained for commercial purposes.

Only one homegarden was included in the minimal management category (type F), and for this reason it was considered an outlier. It was the largest garden of the sample and contained a small number of management zones and high plant species diversity. Two families whose members were mainly involved in carpentry activities jointly managed this homegarden. The labor invested in this homegarden was mostly for controlling fallow vegetation that was prone to fire in the dry season.

### 4.3.4  Discussion and Conclusions

The amount of labor invested in a homegarden was related to family size and the family's dependence on income from the homegarden. No direct relationship between labor investment and the number of zones or plant species existed. The two homegardens with the highest number of zones and plant species (in the type E category) had labor inputs below the sample mean. In these cases, one person worked almost full time in the homegarden. This observation suggests that the quality and consistency of the labor are more influential on homegarden agroecological characteristics than the quantity of work hours invested.

The plant species diversity observed in the sampled homegardens was high compared to other types of agroforestry systems (Nair, 1993). The highest diversity was observed in larger homegardens that produced for household consumption as well as for commercial sales (type E). The different plant use categories and the diversity of species present within these gardens reflected a strategy that sought variety in diet as well as in marketable items (Niñez, 1987).

Dependence on homegarden products as sources of income influenced the type and number of management zones and plant species present in the homegardens. Larger homegardens gave priority to ornamental and coffee zones where families were dependent on homegardens as sources of income (type E). Smaller homegardens that were used to generate income depended mostly on ornamental plant production (type A) which requires less space than that required for other marketable products. In homegardens where products were not used to generate income, fruit trees were the most common plant components (types B and C).

The type of income sources available to a family greatly influenced homegarden management strategies (exemplified here through zonation and plant species selection). In this case study, homegardens represented, in most cases, complementary sources of food and income. Only three of the families were entirely dependent on homegardens for food and income. Because most of the family members engage in low wage labor activities outside the homegardens, this supplementary activity is considerably important for the household. All families considered the homegarden as important space for work and relaxation, and for those involved in handicrafting, it provided space and shade for this activity. Although dependence on homegardens may vary according to specific conditions at a given time (i.e. availability of cash paying jobs), homegardens remain essential resources that are consistently drawn upon to meet the needs of the family.

The homegardens in this study exemplify several of the sustainable characteristics proposed by Torquebiau: uniform and diversified production throughout the year, use of endogenous inputs, and diverse sociological roles. Although some of the other sustainability attributes mentioned by this author were probably present in many of the homegardens, these attributes were not analyzed in this study.

The greatest advantage of this case study is the interdisciplinary nature of the analysis. The data show that homegardens are important agroecosystems that consistently contribute to the household economy and diet while exhibiting many of the agroecological characteristics associated with sustainability that have been reported in other studies. However, this investigation also shared some of the limitations present in other homegarden studies. The study was realized only at one point in time and the agroecological information was incomplete. For example, it did not describe soil characteristics and the potential for fertility conservation and erosion control or analyze perceptions and potential roles of homegardens at the landscape or regional levels. These limitations point to the need for homegarden studies at a higher level of analysis and an increase in the resources available to conduct such research.

## 4.4 SUMMARY AND RESEARCH AGENDA

As local agroecosystems with many demonstrated characteristics of sustainability, tropical homegardens can serve as models, points of reference, and sources of strategies for designing and managing agroforestry systems with a high potential for sustainability. For this reason, homegardens merit further and more careful analysis.

To summarize, homegardens, according to a variety of authors (Gliessman, 1990; Torquebiau, 1992; Jose and Shanmugaratnam, 1993; Gajaseni and Gajaseni, 1999), are potentially important aspects of sustainability and have the following:

- A diversified supply of products and benefits throughout the year
- High plant diversity, mostly for human use, arranged in a structure that is similar to the natural, local forest ecosystem
- Efficient nutrient cycling
- Reduced use of external synthetic inputs
- Social and economic importance for those who invest labor in them
- Management based on a sound, locally developed ecological knowledge base
- Reduced impact on the outside environment

These characteristics fall in line with the definition of sustainable agroecosystems presented in Chapter 1 of this volume. The challenge for future research is to contribute more conclusive data on all these characteristics over longer periods and at different scales. In particular, researchers need to do the following:

- Reach a better understanding of the interaction between the social and agroeco-logical components of homegardens, and the role of this interaction in developing the characteristics of sustainability that have been documented.
- Assess the role of homegardens at the landscape scale (Arnold and Dewees, 1999). Because homegardens are often present in areas with high population densities and increasing urbanization (Landauer and Brazil, 1990; Hoogerbrugge and Fresco, 1993; Lok, 1998a), they may play an important role in regulating the local hydrology and microclimate and in conserving biodiversity (Barry and Rosa, 1996; Collins and Qualset, 1999).
- Analyze the influence of regional social and economic changes on the maintenance and future survival of homegardens.
- Collect quantitative data on the year round production of homegardens in order to better assess the true productive potential of these systems in particular settings.
- Carry out medium to long-term evaluations or time series analyses in order to better assess the sustainable attributes of homegardens over time.

Innovative approaches that integrate a variety of disciplines from the social and the natural sciences will be required to meet the research challenges presented above. Some promising methodologies in this direction can be found in Lok (1998a and b), Rocheleau (1999), Sinclair and Walker (1999), and in the case study presented in this chapter.

# REFERENCES

Altieri, M.A., *Agroecology: The Science of Sustainable Agriculture*, Westview Press, Boulder, CO, 1995.
Altieri, M.A. and Anderson, M.K., An ecological basis for the development of alternative agricultural systems for small farmers in the Third World, *Amer. J. Alternative Agric.*, 1, 30–38, 1986.
Arnold, J.E.M. and Dewees, P.A., Trees in managed landscapes: factors in farmer decision making, in Buck, L.E., Lassoie, J.P., and Fernandes, E.C.M., Eds., *Agroforestry in Sustainable Agricultural Systems*, Lewis Publishers, Boca Raton, FL, 1999, 277–294.
Barry, D. and Rosa, H., Environmental degradation and development options, in Boyce, J.K., Ed., *Economic Policy for Building Peace: The Lessons of El Salvador*, Lynne, Rienner, Boulder, CO, 1996, 233–246.
Collins, W.W. and Qualset, C.O., Eds., *Biodiversity in Agroecosystems*, CRC Press, Boca Raton, FL, 1999.
Ellis, E.C. and Wang, S.M., Sustainable traditional agriculture in the Tai Lake region of China, *Agriculture, Ecosystems and Environment*, 61, 177–193, 1997.
Fernandes, E.C.M. and Nair, P.K.R., An evaluation of the structure and function of tropical homegardens, *Agricultural Syst.*, 21, 279–310, 1986.
Gajaseni, J. and Gajaseni, N., Ecological rationalities of the traditional homegarden system in the Chao Phraya basin, Thailand, *Agroforestry Syst.*, 46, 3–23, 1999.
Gliessman, S.R., Understanding the basis of sustainability for agriculture in the tropics: experiences in Latin America, in Edwards, C.A., Lal, R., Madden, P., Miller, R.H., and House, G., Eds., *Sustainable Agricultural Systems*, Soil and Water Conservation Society, Ankeny, IA, 1990.

Gliessman, S.R., *Agroecology: Ecological Processes in Sustainable Agriculture*, Ann Arbor Press, Ann Arbor, MI, 1998.

Gliessman, S.R., Garcia, R., and Amador, M., The ecological basis for the application of traditional agricultural technology in the management of tropical agro-ecosystems, *Agro-Ecosystems*, 7, 173–185, 1981.

Hoogerbrugge, I.D. and Fresco, L.O., *Homegarden Systems: Agricultural Characteristics and Challenges*, IIED, London, UK, Gatekeeper Series 39, 1993.

Jensen, M., Soil conditions, vegetation structure and biomass of a Javanese homegarden, *Agroforestry Syst.*, 24, 171–186, 1993a.

Jensen, M., Productivity and nutrient cycling in a Javanese homegarden, *Agroforestry Syst.*, 24, 187–201, 1993b.

Jose, D. and Shanmugaratnam, N., Traditional homegardens of Kerala: a sustainable human ecosystem, *Agroforestry Syst.*, 24, 203–213, 1993.

Klee, G., *World Systems of Traditional Resource Management*, Halstead, New York, 1980.

Landauer, K. and Brazil, M., Eds., *Tropical Home Gardens*, United Nations University Press, Tokyo, Japan, 1990.

Lok, R., Estudio de base: San Juan de Oriente y El Castillo, *Informe Interno-Proyecto Huertos Caseros*, CATIE, Turrialba, Costa Rica, 1994.

Lok, R., Ed., Huertos caseros tradicionales de America Central: características, benefícios e importancia, desde un enfoque multidisciplinario, CATIE, Turrialba, Costa Rica, 1998a.

Lok, R., Introducción a los huertos caseros tradicionales tropicales, *Modulo de Enseñanza Agroforestal 3*, Serie Materiales de Enseñanza 41, CATIE, Turrialba, Costa Rica, 1998b.

Madge, C., Ethnography and agroforestry research: a case study from the Gambia, *Agroforestry Syst.*, 32, 127–146, 1995.

Marten, G.G., Ed., *Traditional Agriculture in Southeast Asia: a Human Ecology Perspective*. Westview Press, Boulder, CO, 1986.

Méndez, V. E., Influencia de factores socioeconómicos sobre estructuras agroecológicas de huertos caseros en Nicaragua, M.S. thesis, Tropical Agriculture Research and Education Socioeconómics Center. CATIE, Turrialba, Costa Rica, 1996.

Méndez, V.E., Lok, R., and Somarriba, E., Interdisciplinary analysis of homegardens: a case study from Nicaragua, in Jimenez, F. and Beer, J., Eds., *Proceedings of the International Symposium on Multi-strata Agroforestry Systems with Perennial Crops*, CATIE, Turrialba, Costa Rica, 1999.

Méndez, V.E., Lok, R., and Somarriba, E., Interdisciplinary analysis of homegardens in Nicaragua: zonation, plant use and socioeconomic importance, *Agroforestry Syst.* (in press).

Mergen, F., Research opportunities to improve the production of homegardens, *Agroforestry Syst.*, 5, 57–67, 1987.

Michon, G. and Mary, F., Transforming traditional home gardens and related systems in West Java (Bogor) and West Sumatra (Maninjau), in Landauer, K. and Brazil, M., Eds., *Tropical Home Gardens*, United Nations University Press, Tokyo, Japan, 1990, 169–185.

Nair, P.K.R., Ed., *Agroforestry Systems in the Tropics*, Kluwer Academic Press, Dordrecht, The Netherlands, 1989.

Nair, P.K.R., *An Introduction to Agroforestry*, Kluwer Academic Press, Dordrecht, The Netherlands, 1993.

Nair, P.K.R. 1998. Directions in tropical agroforestry research: past, present, and future, *Agroforestry Systems*, 38, 223–245.

Niñez, V., Household gardens: theoretical and policy considerations, *Agricultural Syst.*, 23, 167–186, 1987.

NRC., *Sustainable Agriculture and the Environment in the Humid Tropics*, National Academy Press, Washington, D.C., 1993.

Rocheleau, D.E., The user perspective and the agroforestry research and action agenda, in Gholz, H.L., Ed., *Agroforestry: Realities, Possibilities and Potentials*, Martinus Nijhoff, Dordrecht, The Netherlands, 1987, 59–88.

Rocheleau, D.E., Confronting complexity, dealing with difference: social context, content and practice in agroforestry, in Buck, L.E., Lassoie, J.P., and Fernandes, E.C.M., Eds., *Agroforestry in Sustainable Agricultural Systems*, Lewis Publishers, Boca Raton, FL, 1999, 191–236.

Sanchez, P.A., Science in agroforestry, *Agroforestry Syst.*, 30, 5–55, 1995.

SAS Institute, *SAS/STAT Guide for Personal Computers: Version 6*, SAS Institute: Cary, NC, 1987.

Scherr, S.J., Economic analysis of agroforestry systems: the farmer's perspective, in Current, D., Lutz, E., and Scherr, S., Eds., *Costs, Benefits, and Farmer Adoption of Agroforestry: Project Experience In Central America and the Caribbean*, World Bank: Washington, D.C., 1995, 28–44.

Schulz, B., Becker, B., and Götsch, E., Locally developed knowledge in a "modern" sustainable agroforestry system: a case study from Eastern Brazil, *Agroforestry Syst.*, 25, 59–69, 1994.

Sinclair, F.L. and Walker, D.H., A utilitarian approach to the incorporation of local knowledge in agroforestry research and extension, in Buck, L.E., Lassoie, J.P., and Fernandes, E.C.M., Eds., *Agroforestry in Sustainable Agricultural Systems*, Lewis Publishers, Boca Raton, FL, 1999, 245–276.

Somarriba, E., ¿Que es agroforesteria? in Jimenez, F. and Vargas, A., Eds., *Apuntes de clase del curso corto: sistemas agroforestales*, CATIE, Turrialba, Costa Rica, 1998, 1–14.

Torquebiau, E., Are tropical agroforestry home gardens sustainable? *Agric., Ecosystems and Environ.*, 41, 189–207, 1992.

Wilken, G.C., *Good farmers: Traditional Agriculture and Resource Management in Mexico and Central America*, University of California Press, Berkeley, CA, 1988.

Wojtkowski, P.A., Toward an understanding of tropical home gardens, *Agroforestry Syst.*, 24, 215–222, 1993.

CHAPTER 5

# Improving Agroecosystem Sustainability Using Organic (Plant-Based) Mulch

Martha E. Rosemeyer

## CONTENTS

1-8493-0894-1/01/$0.00+$.50
© 2001 by CRC Press LLC

## 5.1 INTRODUCTION

Mulches appear so simple that they are often overlooked in discussions of sustainability. Yet the effect of mulches on agroecosystem sustainability is profound, as demonstrated both by research and the practical efforts of organic farmers and managers of traditional agroecosystems.

Mulch is "a layer of dissimilar material separating the soil surface from the atmosphere" (Lal, 1987) or simply a covering applied to the soil surface. Mulches can be made up of a variety of different organic and inorganic substances, including plant material, paper, manure, plastic sheeting, or rock. Organic mulches are often of crop residues, or plants cut and brought in from outside of the cropping system; they may be made up of plants grown *in situ* and cut for mulch, such as the native vegetation of secondary succession, weeds, foliage of alleycropped trees, or green manures. This chapter will focus on the use of organic mulches in the tropics; it will conclude with a discussion of green manures as a special category of mulch.

Mulch systems are ubiquitous in traditional agroecosystems in the humid tropics of both the New and Old Worlds, where they are often mistaken for slash and burn systems (for reviews of mulching see Lal, 1975; Lal, 1977; Thurston, 1997). The use of organic mulches is more common in humid areas because they have sufficient water to produce "fertilizer" crops in addition to food crops (Thurston, 1997). Managers of traditional mulch systems in the tropics use whatever organic materials are available — vegetation cut in swidden systems, crop residues, pruning remains, household refuse, aquatic vegetation cleared from canals, etc.

Open nutrient cycles and simplified food webs are major factors limiting agroecosystem sustainability (Gliessman, 1998). Mulching can address these limitations by preventing erosion and subsequent nutrient loss, increasing internal nutrient cycling, enhancing system biodiversity, and providing (through decomposition) an energy source for the detrital food chain. In addition, mulching generally suppresses weeds, diseases, and pests, reducing or eliminating the need for targeted pest control measures. This latter effect can make a significant contribution to sustainability because herbicides, pesticides, and fungicides represent both an outflow of capital from an agroecosystem and an input of external, nonrenewable energy that can be as high as 67,845 kcal/kg of active ingredient (Fluck, 1995).

## 5.2 THE SLASH MULCH SYSTEM OF TROPICAL CENTRAL AMERICA

This chapter will explore the multifaceted role of mulch in improving the sustainability of agroecosystems. The slash mulch system of tropical Central America will be used to provide specific, well researched examples of the effects of mulch, and these examples will be supplemented by references to work on other systems.

The slash mulch system in the New World was described by early Spanish explorers. Our best documentation of slash mulch systems historically comes from the neotropics, where some of the systems are still in place. Historically the slash mulch system produced beans, maize, sorghum, rice, sugar cane, bananas, and root crops. Slash mulch systems have been described by anthropologists in Africa and

Asia, where maize, beans, sweet potatoes, sesame, sorghum, rice, bananas, and taro were grown (for review of the literature see Thurston, 1997).

Unlike many traditional systems, the slash mulch system is still in wide use in Latin America; today it is particularly relevant to Costa Rican bean production (where the system is called *frijol tapado*). Bean acreage in the slash mulch system has not changed much over the last 20 years and still accounts for 30 to 40% of Costa Rican bean production, 60% of which is sold off the farm (Rosemeyer, 1995). Another system, the unmulched *espequeado*, has been promoted in Costa Rica as a high input, modern system. In the *espequeado* system, the land is cleared, beans are planted with a digging stick, and fertilizers and pesticides are applied (Rosemeyer and Gliessman, 1992).

The slash mulch system is traditionally managed as follows. After about two years of fallow and the selection of an appropriate area based on vegetation, paths are cut in the undergrowth with a machete and seed is broadcast. Then the vegetation between the paths is chopped down and cut up on the ground to form a mulch layer 5 to 20 cm thick. Although this vegetation is often described as containing weeds, it is actually the secondary growth of herbaceous plants and trees. The materials are not weeds in the sense of that weeds represent undesired vegetation. The vegetation is desired for its mulching properties and is not from the common European species found in intensively cropped systems in the New World. The bean seeds germinate and emerge from the mulch layer. Essentially no cultural practices are employed until harvest.

This system is considered sustainable because it has been practiced for centuries with no apparent negative environmental effects (Thurston, 1997). The key factor in its environmental sustainability is the mulch layer of secondary growth vegetation. This layer contributes to the closure of the nutrient cycle by promoting high internal nutrient cycling and enhancing the complexity of the energy flow of the system — major factors for sustainability.

Moreover, the mulch layer helps the slash mulch system resemble the natural ecosystems in the region (Bunch, 1995). The root systems of the growing beans ramify into the mulch layer, forming a root mulch structure containing the bulk of the bean plants' root systems (Figure 5.1). Researchers estimate that between 60% (Woike, 1997) and 85% (Woike and Rosemeyer, in preparation) of the roots are in the mulch, rather than in the soil. This root mulch structure is similar to the root litter mats common in natural tropical forest systems, particularly those with poor soil (Jordan, 1985). In the Venezuelan Amazon, for example, 20 to 25% of root biomass is in the root litter mat (Jordan, 1985); the majority of the more functionally important and nearly weightless feeder roots involved in uptake can be in this layer. Nutrient cycling is considered direct because upon decomposition few of the nutrients escape from immediate plant uptake. Root litter mats are so effective that 99.9% of radioactive Ca and P applied to the mat was retrieved by roots and only 0.1% leached. Sixty to 80% of other nutrient cations are retrieved by the root litter mat (Stark and Jordan, 1978).

The slash mulch system bears another important resemblance to local natural systems: the diversity in the system imitates the natural method of energy capture and nutrient cycling, restores native fertility, and helps control pests and disease, thereby alleviating the necessity for agrochemicals. The slash mulch system fosters biodiversity by using mulch of native successional vegetation which is growing for

**Figure 5.1** In the slash mulch system (left), the majority of each crop plant's roots are in the mulch layer, facilitating internal nutrient cycling and limiting leaching losses. In unmulched systems (right), the roots are restricted to the soil.

at least 9 months between bean planting seasons. Fifty or more species were commonly found in the second growth vegetation used for mulch; traditional farmers choose sites for the slash mulch system based on the species composition of the vegetation (Meléndez et al., 1999; Kettler, 1996; Araya and Gonzalez, 1986).

In the last few decades, demand for higher production has compressed the traditional 2 to 4 year fallow period to as short as 9 months, and beans are produced every year (Bellows, 1992; Rosemeyer et al., 1999a). Consequently, the secondary vegetation is degraded and dicotyledonous plants replaced with monocots that are less productive for slash mulch beans. To compensate for the shorter fallow periods, some slash mulch system managers have been planting alley cropped leguminous trees, which are coppiced yearly so that their high quality foliage can be used for mulch. In a series of experiments spanning more than a decade, agronomic and nutrient cycling aspects of the slash mulch system, the modified alley crop mulch system, and unmulched systems have been explored (Rosemeyer and Gliessman, 1992; Rosemeyer, 1994; Kettler, 1997a; Schlather, 1998; Rosemeyer et al., 1999a; Melendez et al., 1999; Rosemeyer et al., in press).

## 5.3 EFFECTS OF MULCH

### 5.3.1 Erosion Control

Unsustainable rates of soil erosion represent a major threat to agroecosystems worldwide. Soil degradation due to water erosion affects 55.6% of the world's agricultural lands to varying degrees (Oldeman et al., 1991). On sloping agricultural land, which is responsible for producing the majority of local foodstuffs in Latin America (Posner, 1982) and a significant percentage of export production, particularly coffee (Rosemeyer et al., 1999a), average erosion rates for annual cropping are greater than 100 t/ha/yr and can reach 289 t/ha/yr on the steeper slopes (Solórzano et al., 1991). In contrast, the estimated renewal rate for these soils is about 1 t/ha/yr (Pimentel, 1993). This problem is serious because eroded soil is the result of a selective process and contains higher quantities of nutrients and organic matter than the rest of the soil (El-Swaify, 1993).

Mulching, however, keeps the soil in place and prevents the nutrient and organic matter losses associated with erosion. Surface soil erosion is reduced in proportion to the depth of the soil surface cover; a good soil cover is the most effective line of defense against surface and gully erosion due to water (Hamilton, 1994).

In slash mulch bean production systems on hillsides in Costa Rica, erosion was 6 times less than that in similar systems with bare soil (Bellows, 1992). When vegetation was burned instead of mulched, soil loss increased 8 times in Indonesia (Lal, 1996). In Korea, the inclusion of a mulch of soybean residues on slopes as steep as 15% decreased surface erosion 86 to 90% compared to slopes on which conventional tillage was used (Lal, 1996). In the Philippines, erosion was decreased 65% by the use of vegetative barriers, but it was decreased 95% by mulching (Garrity, 1993). In the Philippines, soil loss on 14 to 21% slopes was reduced from 105 t/ha to only 5 t/ha by alley cropping and mulching with tree prunings and crop residues (Griggs, 1995). Similar examples of the effects of residue mulch systems on erosion in other soils and ecosystems in the tropics are reviewed in Lal (1990) and Thurston (1997).

### 5.3.2  Increased Internal Cycling of Nutrients

When mulch is comprised of vegetation grown *in situ*, as is the case with the slash mulch system, the nutrients contained in the mulch biomass remain in the system, facilitating internal nutrient cycling. Moreover, maintaining mulch on the soil, as opposed to burning it, allows the nutrients in the organic residue to be more accessible to the crop plants. In slash and burn systems, nitrogen, carbon, and sulfur are rapidly volatilized during burning; loss of these elements has been measured at 30%, 20%, and 49% respectively (Ewel et al., 1981). Slash mulch systems, in contrast, store nutrients in decaying organic matter on top of the soil, where they are taken up efficiently by the plant roots occupying the mulch layer.

The amount of nutrients contained in mulch can be considerable. At one site, the quantity of mulch applied in a slash mulch bean system was estimated to be between 10 to 30 t/ha/year of second growth vegetation and weeds, which is equivalent to the application of 154 to 450 kg N/ha and 11 to 33 kg P/ha (Rosemeyer and Kettler, in preparation). At another site, an estimated 5 t/ha of biomass was applied, which represents 53 kg N/ha and 9 kg P/ha (Meléndez and Szott, 1999). When the fallow of the slash mulch system was enriched with tree prunings in an alley cropped system, 6 years after treatment implementation, 30 t/ha/yr (dry matter) was applied and found equivalent to 300 kg N/ha and 20 kg P/ha (Rosemeyer and Kettler, in preparation). Haggar (1990) found that 10% of the N from mulch of slashed and mulched alley cropped tree foliage was available to the crop plant (maize) during the growth cycle. Even if only 10% of the nutrients stored in mulch decompose and are available during the bean growing season, this represents a substantial input of available nutrients.

Fertilizer replacement values of the slash mulch in three successive years of bean cropping were found to be equivalent to the soil application of 43 kg inorganic P/ha (Rosemeyer, 1996). Levels of available P in the soil were significantly greater in the mulch system than in the unmulched system at both 0 to 5 cm and 5 to 10 cm soil depths, and averaged 39 to 41 ppm P in the mulch system versus 36 ppm P in the unmulched. The water fraction of the litter layer of the mulched and unmulched systems contained 1.5 kg P/ha and 0.3 kg P/ha, respectively (Schlather, 1995).

The improved nutrient cycling dynamics observed in the slash mulch system applies generally to other systems. In Argentina, for example, P was increased in all soil fractions in surface and subsoil when an elephant grass mulch was applied to a perennial crop (*Ilex paraguariensis*), increasing P sustainability of the system (Camelo et al., 1996).

### 5.3.3  Increased Efficiency of Applied Inorganic Nutrients in Augmenting Yields

Increased efficiency of applied nutrients — that is, greater uptake of the applied nutrients by the crop plants — is a goal of sustainable agricultural systems, especially when the applied nutrients are exogenous and from nonrenewable sources (Gliessman, 1998). Mulches may contribute to this goal by improving crop utilization of applied inorganic nutrients. In African coffee systems, for example, farmers and researchers have observed that when fertilizer is mixed with grass and the mixture applied as mulch, the system is more profitable than when fertilizer is applied directly to the soil (Wellman, 1961). Similarly, applying inorganic fertilizer along with litter, manure, and termitarium soil (from termite mounds) has been shown to increase the yield response of corn compared to application of fertilizer alone (Campbell et al., 1998). Such effects may be explained by the ability of the added organic material to increase soil moisture, increase cation exchange capacity (CEC), and provide complementary nutrients lacking or at ineffective proportions in the fertilizer. In addition, nutrient losses are minimized when organic materials are applied with inorganic fertilizers. When urea is used as a fertilizer, for example, mulches can prevent volatilization of ammonia (Campbell et al., 1998).

Tests of fertilizer application in the slash mulch bean system showed increased efficiency of applied nutrients. The ratio of bean yield to applied P was higher in plots using the traditional mulching techniques than in plots without mulch (Figure 5.2). Since P is the limiting nutrient (Rosemeyer and Gliessman 1992) in the system, the observed differences were probably due to increased P availability in the mulched plots. Several mechanisms may account for this increase in P availability:

1. In the unmulched system, P was immobilized in the soil. The Andisol soil type fixed approximately 86% of the P, resulting in low availability of applied P (Rosemeyer, 1990).
2. The bean roots in the mulched system, most of which are located in the mulch, are able to take up nutrients more quickly and directly than the roots in the unmulched system, which are restricted to the soil.
3. The decomposing mulch creates a pH that is more conducive to plant uptake of nutrients than the pH in the unmulched soil. The pH of soil under the decomposing mulch was found to be higher than that of the soil in the unmulched system (Mata et al., 1999).

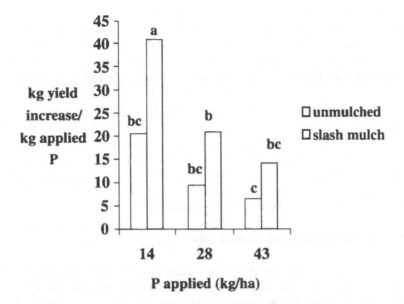

**Figure 5.2** Efficiency of three levels of phosphorus applied to mulched and unmulched systems at Finca Loma Linda, Costa Rica, averaged from data collected from 1992 to 1995. Bars labeled with the same letters represent values that do not differ significantly according to Duncan's multiple range test at the 5% level. (Adapted from Schlather and Rosemeyer, in preparation.)

Other studies have found no evidence that added fertilizer P is adsorbed less strongly by soil particles beneath the mulch in alley cropped soils than in unmulched systems (Haggar, 1990), leading one to think that the dynamics of P in the decomposing mulch solution must be affected. This corroborates farmer observation. Producers in Jamaica say that nutrients for plant uptake come from the rotting vegetation, not the soil (Thurston, 1997).

### 5.3.4 Moisture Retention

Mulch material placed at the soil surface reduces evaporation by protecting the moist layer of air close to the surface from wind and by reducing soil temperature. Mulches have the effect of lowering the maximum soil temperature because they generally reflect more and absorb less solar radiation and have lower thermal conductivity than soil (Jalota and Prihar, 1998). The insulation of the ground from air temperature and radiation depends on the thickness of the mulch layer; for example, 8 to 13 t/ha of straw mulch resulted in a lower ground temperature during a hot period and higher soil temperature during a cold period than 4 t/ha of straw mulch (Unger, 1978). Low temperature at the soil surface underneath the mulch lowers the vapor pressure of the soil surface and consequently the vapor pressure gradient between the soil surface and the mulch atmosphere above it. Mulch also provides a barrier for water movement to the atmosphere.

Overall reduction of evaporative water loss with mulch is influenced by soil type, evaporativity (initial potential for evaporation), the nature and amount of residue, timing and manner of mulch placement, precipitation patterns, and other climate and tillage factors (Jalota and Prihar, 1998). In general, increases in soil water with mulch depend on the amount of mulched material, although the water storage efficiency *per unit* of mulch decreases slightly with increase in mulch rate.

In the slash mulch system, which is generally practiced where farmers are unable to burn due to excessive humidity (Thurston, 1997), beans can be more susceptible to drought stress because the majority of the root system is in the mulch, not in the soil. Nevertheless, the mulch generally keeps the soil under it more humid under dry conditions (Rosemeyer, unpublished data). This relationship shows that when analyzing the interaction of mulch systems and environment, the location of the rooting is critical to understanding system function.

### 5.3.5  Promotion of Root Symbioses

Mulch can hypothetically provide a more stable microenvironment that facilitates nodulation and mycorrhizal colonization, and permit greater extraction of nutrients from low external input agroecosystems (Rosemeyer and Gliessman, 1992). Reports of the effect of mulches on nodulation in the literature are mixed. In Malaysia, Masefield (1957) found that grass clippings increased the nodulation of cowpeas threefold, but in Brazil, dry grass mulch did not significantly effect nodulation of the common bean (Ramos and Boddey, 1987). With respect to mycorrhizae, reports are scant. In no-tillage production in the Netherlands utilizing mulching, mycorrhizal fungus infection was greater than in a conventional, plowed system (Ruissen, 1982).

Evidence from the slash mulch system points to promotion of plant symbioses in the soil by the mulch under certain conditions. The biomass of nodules per plant was greater on bean roots in the slash mulch plots than on bean roots in the unmulched plots in 2 of 3 years when growing conditions were relatively dry. However, in only one of those 2 years was the difference significant, due to the variability associated with nodulation (Rosemeyer et al., in press).

Experiments with different types and quantities of mulch vegetation in the slash mulch system show that some types of mulch can reduce bean nodulation. In the alley cropping enrichment experiments, the nodulation of beans under a mulch enriched with *Calliandra calothrysus* was significantly less than that of an unmulched treatment in all 3 weeks of measurement (Rosemeyer et al., in press). When different alleycrop mulches are compared in orthogonal contrasts, nodulation is depressed in beans grown under mulches enriched with *Calliandra calothrysus* and *Inga edulis* relative to mulches enriched with *Gliricidia sepium* and normal slash mulch at both 3 and 5 weeks after bean planting (Figure 5.3). The reduction of nodulation using *Calliandra* and *Inga* mulches may be due to the high quantities of N released from the decomposing vegetation in these two treatments (a hypothesis that is presently being tested). High quantities of applied N are known to depress nodulation (Sprent and Minchin, 1983). With more sampling, we may see the depressive effect decrease over time with *Calliandra* but not *Inga*. This may correlate with *Calliandra*'s faster rate of decomposition and amounts of N released (Kettler, 1997b).

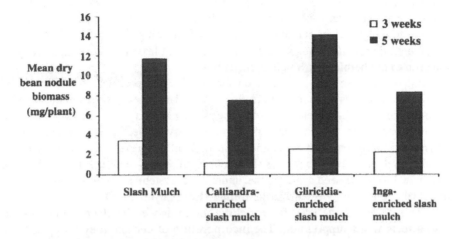

**Figure 5.3**   The nodulation of beans under four different mulches at Finca Loma Linda, Costa Rica, 1997. The latter three mulches were each enriched with vegetation from alley cropped trees (*Calliandra calothrysus, Gliricidia sepium,* and *Inga edulis,* respectively). Beans were sampled 3 and 5 weeks after planting.

In order to assess the role of microorganisms in maintaining soil or system health, root mutualistic symbioses (mycorrhizae and legume nodulation) should be examined carefully, especially because of their important contributions under conditions of low nutrient availability. Symbioses may be depressed by high nutrient levels, either exogenous, in the case of inorganic fertilizers, or endogenous, in the case of certain quantities or types of mulch (Rosemeyer et al., in press).

## 5.3.6   Weed Supression

One of the most important effects of mulch is weed suppression during crop growth. Traditionally no labor was needed for weed control in mulch systems (Rosemeyer, 1995), although the decrease in fallow time in the slash mulch system has made some weeding necessary. Slash mulch farmers typically spend about half as much time in weed control as do farmers using the unmulched system (Rumoroso and Torres, 1999). For this reason, the mulch system decreases the need for herbicide, an input with a high nonrenewable energy content.

Examples of weed suppression by mulch are abundant in the literature. Compared to an unmulched control, weeds were reduced by 57% and rubber seedling growth enhanced significantly with a mulch of plant material (Lakshmanan et al., 1995). In India, a 7.5-cm layer of coir pith (fibrous coconut seed mesocarp) used as a soil mulch for cashews decreased weed growth 73% in comparison to the unmulched control (Kumar et al., 1989). In Antigua, West Indies, dried Guinea grass mulch, applied at rates of 4 and 8 t/ha on cowpeas and eggplants, reduced weed growth more effectively than an unmulched system, and increased water retention and crop seedling germination (Daisley et al., 1988). In India, organic mulch distillation waste of citronella Java (*Cymbopogon winterianus*) applied at the rate of 3 t/ha was more effective than three herbicides in control of weeds in lemon grass (*C. flexuosus*) and two other aromatic grasses (Singh et al., 1991). Also in India, organic mulch was superior to six herbicides at reducing weeds and increasing yields of medicinal yams (*Dioscorea floribunda*). Yields were increased due, at least in part, to sensitivity of the crop to the herbicides (Singh et al., 1986).

Mulch may also have an affect on the species composition of weeds. In Sri Lanka, weed species were reduced from 11 to 5 when pineapple was mulched with coconut coir dust (Mele et al., 1996). The slash mulch system favors weeds that resprout from roots, while the unmulched system favors weeds that start from seeds (Rosemeyer and Kettler, in preparation). Several authors have noted a similar effect in other systems (Budelman, 1988; Ikuenobe et al., 1994). Fewer grass seeds were found in the weed seed bank in the unmulched system in comparison with the mulched due to hand weeding in the former (Rosemeyer, 1995).

The type of foliage used for enriching the mulch in the slash mulch system also affects weed suppression. The incorporation of certain alley cropped trees with slowly decomposing foliage (e.g., *Inga edulis*) into the system suppresses growth of weed biomass more effectively than other trees (e.g., *Gliricidia sepium*) (Rosemeyer and Kettler, in preparation). Similarly, in the Ivory Coast, foliage of the nitrogen fixing tree *Fleminigia macrophylla* was superior to that of *Gliricidium sepium* and *Leucaena leucocephala* in suppressing weeds that multiply by seeds (Budelman, 1988). In Nigeria, weed control during the corn cropping season was more effective in alley crop derived mulches of *Cassia* than it was in mulches derived from *Gliricida* and *Flemingia* (Yamaoh et al., 1986). Based on data from Africa, the estimated labor requirement for hypothetical alley cropped systems was 460 hours/ha for *Leucaena*, 108 for *Gliricidia*, and 23 for *Fleminigia*, with weed dry matter reduced 53%, 64%, and 92%, respectively (Bohringer, 1991).

### 5.3.7  Disease Suppression

Mulch and alleycropping systems commonly suppress plant pathogens (Rosemeyer et al., in press), especially fungal pathogens, possibly because the mulch provides a physical barrier, changes the physical environment, or intensifies microbial activity. However, mulches can also provide habitats in which some pathogens can feed and reproduce (Thurston, 1997).

The slash mulch system has been found to suppress several diseases of beans — an important effect in light of the fact that diseases are the most important limiting factors in bean production in Costa Rica (Arias and Amador, 1990). Galindo et al. (1983) found web blight of beans (*Thanatephorus cucumeris*, sexual stage; *Rhizoctonia solani*, asexual stage) suppressed in the slash mulch system or with rice hull mulch. It is hypothesized that the physical barrier of the mulch prevents the splashing of soil-borne sclerotia and thick walled hyphae onto foliage (Galindo et al., 1983). Rain splash is the second most important natural agent after wind in the dispersion of spores of plant pathogenic fungi (Fitt and McCartney, 1986). Additionally, microbial activity in the mulch might suppress or inhibit the raindrop splashed inoculum from reaching the leaves.

The decomposing slash mulch vegetation demonstrated 5 times greater microbial respiration than the litter or the soil. High microbial activity associated with plant decomposition may provide a barrier of actively metabolizing microbes and associated soil fauna that may be antagonistic to the spores of a fungal plant pathogen. In other words, the mulch barrier is not only a physical barrier to plant pathogens and an important site of decomposition and nutrient cycling, it may also be a source of general microbial activity that can prevent the establishment of any one pathogenic microorganism (Rosemeyer et al., in press).

Anthracnose (*Colletotrichum lindemuthianum*) — the most serious foliar disease of beans in the world (Pastor-Corrales and Tu, 1989) — was significantly reduced under the slash mulch system. There was a higher incidence of anthracnose affected leaves in the unmulched plots than in the slash mulch treatments, but not all mulch species affected disease incidence similarly (Table 5.1). Mulch of a second growth plant thought to be favorable for the slash mulch system, *Melanthera aspera* (Kettler, 1996; Melendez et al., 1999), resulted in significantly more anthracnose on bean foliage than noted with mulch of *Mucuna* spp. (Table 5.1).

Similarly, *Fusarium* root rot disease incidence and severity were lower in slash mulch and alley cropped systems than in unmulched systems (Table 5.1). *Fusarium* chlamydospores in soil or infected plant residue are stimulated to germinate by nearby bean seed or root exudates (Abawi, 1989). Since the bean seed germinates in the mulch (which does not usually include bean plant residue) and the majority of the bean root system remains in the mulch layer, as opposed to the soil (Rosemeyer and Woike, unpublished), the bean plant in this system is essentially avoiding the source of inoculum by proliferating in the decomposing mulch layer. This observation suggests that the phrase "soil health" should be replaced with "system health," since the bean plants in the slash mulch system exhibit fewer disease symptoms by avoiding the soil.

The slash mulch system does not appear to suppress all plant pathogens. Damage due to *Rhizoctonia* root rot was higher in the slash mulch and alley crop systems

Table 5.1   Incidence of Bean Diseases in Slash Mulched and Unmulched Systems at Finca Loma Linda, Costa Rica, 1994–1995

| System | Anthracnose Colletotrichum | Root rot Fusarium | Root rot Rhizoctonia |
|---|---|---|---|
| Slash mulch | 2.6 | 2.9 | 1.8 |
| *Melanthera aspera* enriched mulch | 4.1 | 2.5 | 2.1 |
| *Mucuna* spp. enriched mulch | 2.2 | 3.1 | 2.4 |
| Unmulched | 5.25 | 3.9 | 1.4 |
| F orthogonal contrast | 45.6[a] | 7.40[b] | 6.70[c] |

Measured by the CIAT 1 (low) to 9 (high) scale (anthracnose and *Rhizoctonia* root rot) and the Abawi 1 (low) to 9 (high) scale (*Fusarium* root rot; Abawi, personal communication). Orthogonal contrasts between the slash mulched and unmulched bean means (of unfertilized and fertilizer treatments) were significantly different for all three diseases.

[a] P <0.001.
[b] P <0.01.
[c] P <0.05.

than in the unmulched systems (Table 5.1). The effect of this pathogen on yield was not directly measured, although observed yields were generally greater in the slash mulch or alley crop mulch systems than the unmulched system. Manning et al. (1967) found that deep planting of beans in soil favors *Rhizoctonia* infection, suggesting that the greater the length of the hypocotyl, the greater the chance that seedling tissue will be exposed to the pathogen. Hypocotyl length was significantly greater in the slash mulch and alley crop mulch plots than in the unmulched plots (Rosemeyer et al., in press). In general, increased incidence of *Rhizoctonia* has been associated with no-till beans (Abawi and Pastor-Corrales, 1990) and no-till systems in general (Abawi and Thurston, 1994; Pankhurst, 1994), probably due to greater contact between the bean hypocotyl and infected residues from previous crops and weeds. Since *Rhizoctonia solani* is found worldwide and in uncultivated soils, weeds and native vegetation may be involved (Baker and Martinson, 1970).

Mulches of various types affected disease incidence differently in our studies. For example, *Melanthera* mulch resulted in significantly more anthracnose than *Mucuna* mulch or mixed species mulch (Rosemeyer et al., in press). This result suggests that further research in this area may help farmers control a particular disease by planting alley cropping species shown to be most effective in control of that disease, or by avoiding alley cropping species shown to be connected with a higher incidence of a disease. In a preliminary experiment, mulch of bean residues and mulch of a grass (*Melinis minutiflora*) increased angular leaf spot disease (*Phaeoisariopsis griseola*) more than mulches of two grasses, a dicot, and a fern, probably due to infection from the residues (Rosemeyer, 1985).

Disease suppression by mulch has been reported in a number of tropical crops (Thurston, 1992). In Malawi, tomato diseases were reduced along with insect pests and sunburned fruit with a 10 cm layer of barnyard grass mulch (Kwapata, 1991). Mulching of cassava reduced stem tip dieback of unknown etiology in Zaire (Muimba-Konkolongo et al., 1989). In Kenya, black rot of cabbage (*Xanthamonas campestris* pv. *campestris*) was controlled with grass mulch applied immediately after transplanting, and its effect was equal to that of a copper based fungicide with bacteriocidal properties (Onsondo, 1987). In a conservation tillage experiment in Mexico involving herbicides and mulching of corn over 6 years, no severe weed, insect, or disease problems arose (Palmer, 1985). However, rice hulls, cocoa leaves, and sawdust did not decrease *Phytopthora* disease in cacao in Honduras (Porras and Sanchez, 1991), and cassava peel mulch increased fungal disease of tomato and eggplant in Nigeria (Asiegbu, 1991).

We can conclude that mulch use generally reduces the need for fungicides that can impact human health, reduce mycorrhizae, decrease litter decomposition, and impact nutrient cycling. For these reasons, and because fungicides are estimated to represent significant external energy inputs, the disease suppression effect of mulch can be said to make an important contribution to sustainability.

### 5.3.8  Changes in Pest–Crop Interaction

Positive effects of the slash mulch system on system biodiversity and microhabitat heterogeneity should encourage natural control of pest populations (Gliessman,

1998). Studies of the slash mulch system appear to support this hypothesis. Beans grown in the slash mulch system are less susceptible to damage from several insect pests than beans grown in unmulched systems (Arias and Amador, 1990).

Chrysomelid beetles (*Diabrotica* spp.) are considered the second most important insect pests of beans in Latin America (Cardona, 1989). The greatest damage is caused by the adult beetles, which can consume a large portion of a plant when it is a seedling. The adult beetles also can transmit various virus diseases such as bean rugose mosaic virus (BRMV) (Cardona, 1989). The beetles are considered a greater problem in unmulched systems than in mulched systems (Arias and Amador, 1990). Of farmers using the unmulched system in the bean producing Petiballe region of southern Costa Rica 14 to 25% reported using an insecticide for Chrysomelids, while no mulch using farmers did so. However, it should be noted that when the cover crop plant *Mucuna* was used to enrich the fallow, bean seedling germination and emergence were reduced relative to the normal slash mulch system due to larval Chryosmelid damage (Schlather and Rosemeyer, unpublished data).

One important pest in Central America, the aggressive slug *Sarasinula plebia*, is increased by mulching. This pest was introduced into Central America relatively recently, and has been a problem pest of beans since 1967 (Hallman and Andrews, 1989). In the Petiballe region, 51% of bean farmers using the unmulched system and 44% of the slash mulch bean farmers reported the slug to be a problem. Molluscicides are applied by 35% of the farmers at 10 to 25% of the recommended rates. One factor contributing to the slug problem may be the shortening of fallow periods in the slash mulch system; longer fallow periods have been shown to decrease slug problems (Bellows, 1992).

Research on other systems shows mulches to reduce the damage caused by certain pests. In intercropped sorghum and maize in Honduras, a slash mulch system had a significantly lower infestation of fall army worm (*Spodoptera frugiperda*) on whorl stage maize than did a slash and burn system, with similar numbers on plants at time of peak infestation. In burned fields, the neotropical stalk borer (*Diatrea lineolata*) damaged sorghum more than it did in mulched fields (Castro et al., 1998). In Chad, mulching of cassava positively affected the biological control of cassava mealybug by a parasitoid (Neuenschwander, 1990). In Venezuela, bush tomato plants mulched with six different organic mulches and black polyethylene had less damage from the pest *Neolkeucinodes elegantalis* than did unmulched plants (Aponte et al., 1992). In Taiwan, rice straw or plastic mulch reduced the sweet potato weevil infestation in sweet potato more than flooding alone (Talekar, 1987).

The more mulch aids in the regulation of pest populations and thus reduces the use of insecticide, the greater its importance in increasing the sustainability of agroecosystems. Insecticide use can impact human health, cause secondary pest resurgence, reduce populations of natural enemies and soil insects, and represents a significant inflow of nonrenewable energy.

## 5.3.9 Soil Biodiversity Enhancement

Decomposing organic mulch is an energy source for the below ground food web, an assemblage of organisms that facilitates decomposition and therefore plays an important role in internal nutrient cycling. The below ground food web differs from the above ground food web in that its foundation is heterotrophic instead of autotrophic, but both food webs have primary, secondary, and tertiary consumers. The primary consumers of organic matter are the microbes, bacteria, fungi and actinomycetes, whose important role in decomposition is facilitated by comminution of the soil detritivores. The second order consumers include bactivorous and fungivorous nematodes, fungivorous Collembola, and protozoa. Tertiary consumers include some of the larger predatory nematodes, mites, and beetles. There is evidence that a larger energy source in the form of mulch increases the abundance of soil biota, and we may also hypothesize that the larger energy source increases biodiversity within the soil food web.

The abundance and diversity of soil macrofauna in the slash mulch, alley crop mulch, and unmulched systems were compared during the fallow period between bean growing seasons. The slash mulch and alley crop mulch systems showed significantly greater arthropod abundance and morphospecies richness than the unmulched system (Rosemeyer et al., 1999b) (Figures 5.4 and 5.5). Similarly, a greater abundance of detritivores was observed in the slash mulch system (an average of 8.5 detrivores per pitfall trap) than in the unmulched system (an average of 2.75 per trap) (Cook et al., 1993).

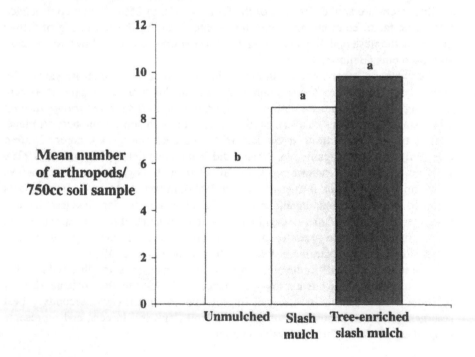

**Figure 5.4** Arthropod abundance in unmulched, slash mulched, and tree-enriched slash mulched plots, Finca Loma Linda, Costa Rica, 1998. Bars marked with the same letters represent values that do not differ significantly according to Duncan's multiple range test, P <0.05. (Adapted from McGlynn et al., in preparation.)

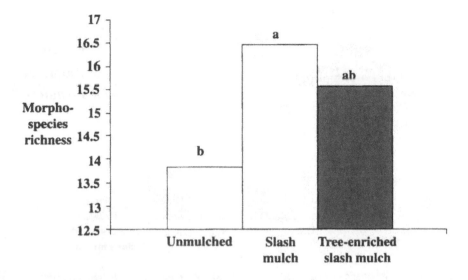

**Figure 5.5**   Arthropod morphospecies richness in unmulched, slash mulched, and tree-enriched slash mulched plots, Finca Loma Linda, Costa Rica, 1998. Bars marked with the same letters represent values that do not differ significantly according to Duncan's multiple range test, P <0.05. (Adapted from McGlynn et al., in preparation.)

In a semi arid environment in central Queensland, the effects of modified tillage practices on populations of soil macrofauna were investigated. Zero tillage systems had the highest abundance of macrofauna species ($74/m^2$), followed by reduced tillage, stubble mulch, and conventional tillage systems (an average of $31/m^2$); the zero tillage systems had the greatest diversity of macrofauna species (Wilson-Rummenie et al., 1999). In another study, mulches of five tropical plant species were compared, along with an unmulched control, for their effects on abundance of soil macrofauna. Relative to the control, the grouped mulch treatments increased earthworm populations by 41% and termite abundance by 177% (Tian et al., 1993).

Abundance of certain biota is affected by the species of plant that makes up the mulch. In the above mentioned study of five tropical species used as mulches, ant populations were 36% higher with *Leucaena* and *Gliricidia* prunings compared to the control, and were similar to the control with mulches of *Acioa*, maize stover, and rice straw. These results showed that ant populations were significantly correlated to the N content of the plant material; similarly, earthworm populations were negatively affected by the ratio of lignin to nitrogen (Tian et al., 1993). In an alley cropping system in Nigeria, soil macrofauna contributed to 30 to 40% of mulch decomposition during the first half life of a mulch of *Flemingia congesta* and *Dactyladenia barteri*. After a longer decomposition period, this faunal effect became more apparent in the more slowly decomposing mulch of *Dactyladenia* (Henrot and Brusaard, 1997). Both of these studies imply that the chemical composition of the plant material used for mulch plays a critical role in soil faunal abundance through its effects on palatability and decomposability.

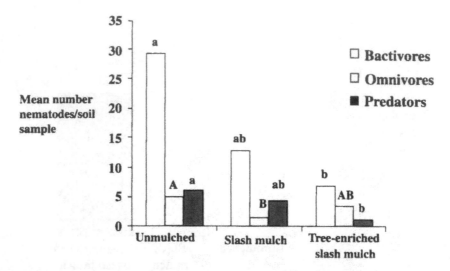

**Figure 5.6**  Abundance of nematodes in three trophic groups in unmulched, traditional slash mulched, and tree enriched slash mulched systems, Finca Loma Linda, Costa Rica, 1998. Bars marked with the same letters represent values that do not differ significantly according to Duncan's multiple range test, P <0.05.

Preliminary data from the slash mulch system indicates that mulch has an effect on the populations of soil nematodes, an important component of the below ground food web (see Chapter 7 in this volume). This effect is complex; the unmulched, traditional slash mulched, and alley crop mulched systems have significantly different proportions of the three different functional or trophic groups of nematodes (Figure 5.6). In general, nematodes are more equally distributed among the three trophic groups in the two mulched systems than they are in the unmulched system. The alley crop mulched plots had fewer predatory nematodes and bactivores than did the unmulched plots. Bactivore diversity and abundance were greatest in the unmulched plots. Bactivorous nematodes often appear in relatively high numbers in unmulched, conventionally managed cropping systems because of their tendency toward a bacterially based food web (Coleman and Crossley, 1996). In slash mulched plots, the abundance of bactivores and predatory nematodes was intermediate and not distinguishable. However, fewer omnivores were present in the slash mulch soil than were in the unmulched soil (Rosemeyer et al., 1999b). In New Zealand, experiments comparing weed control practices in corn and asparagus cropping systems found that sawdust mulch, as compared to hoeing, cultivation, and herbicides, had the greatest positive effect on populations of bactivorous and fungivorous nematodes, but the "increase in populations of predaceous nematodes may be responsible for the absence of marked increases in other functional groups" (Yeates et al., 1993). These differences in nematode community composition have an unknown impact on nutrient cycling function.

### 5.3.10 Reduction of Human Labor

The reduction of human labor involved in mulch systems is substantial, not only in weed control but also in pest control. The labor needed per hectare is 30 to 33 person

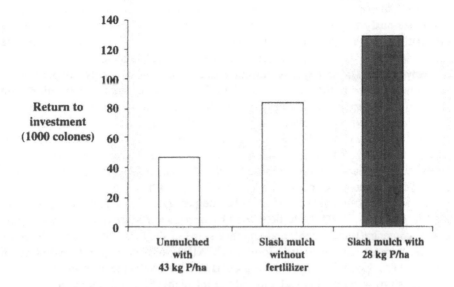

**Figure 5.7**   Economic return on financial investment for the slash mulched, fertilized unmulched, and fertilized slash mulched agroecosystems, Finca Loma Linda, Costa Rica, 1990 to 1992. Seed (usually saved from previous seasons) and labor costs were not included in the assessment.

days for the slash mulch system and 36 person days for the unmulched system, which additionally requires pesticides and herbicides to remain viable. The benefit to cost ratio for the slash mulch system is 1.7 (unfertilized) to 3.13 (fertilized with 28 kg rock P/ha) in comparison to 0.96 for the unmulched system and 1.43 for a semi-mechanized system (Rumoroso and Torres, 1999). The risk to investment was also found to be less with the slash mulch system compared to the unmulched system (Arias et al., 1999). When 28 kg/ha of inorganic P fertilizer (annual average) was applied to the slash mulch system, the economic return to financial investment was 60% more than the return of the unfertilized slash mulch system and nearly 200% more than the return of the unmulched system (Figure 5.7). The financial gain per hour of labor was also found to be more favorable with the slash mulch system than with other systems of bean production (Ribier et al., 1988). Since the system requires little labor, in areas where the bean planting season coincides with high demand for more remunerative labor, (e.g., during the coffee harvest), the system is also favored over other more labor intensive systems of bean production (Rosemeyer, 1995; Bellows, 1992).

## 5.4 FUTURE OF MULCH SYSTEMS

With increasing intensity of land use, mulches of natural vegetation (such as those used in the slash mulch system) are gradually being replaced with mulches of cultivated plants, particularly herbaceous and arboreous nitrogen-fixers. This type of system, in which a crop is planted during a managed or enriched fallow and its biomass left in the system as a living or dead mulch, may use mulch producing plants variously termed green manures, cover crops, or smother crops, depending on their primary role. More closely managing the fallow in this way will become more compelling as pressure on the land base continues to increase.

Velvetbean (*Mucuna sp.*) is currently one of the most widely adopted cover crops/mulch crops in the world. More than 50 nongovernmental organizations and research groups promote this plant or conduct research on it (Buckles, 1994). A current article reviews the history of the use of this plant and how it has waxed and waned over the world, with many thousands of farmers in Latin America and Africa using it to raise yields and control weeds (Buckles et al., 1998). Vines can grow up to 15 m and produce biomass of up to 50 t/ha (Bunch, 1994). Where the velvetbean is used in Central America, maize yields of 3 to 4 t/ha on hillsides are common (Triomphe 1996), and in West Africa it has helped raise yields 10-fold, from 200 to 2000 kg/ha (IITA, 1993). It has been used by farmers to restore tilth and soil organic matter contents (Bunch 1990) and smother some of the most difficult weeds in the tropics, *Imperata contracta* in Colombia (CONIF, 1987) and *Imperata cylindrica* in Africa (IITA, 1993). The restorative properties of the velvetbean have reduced the need for shifting cultivation and allowed the intensification of the corn–bean system on the north coast of Honduras (Buckles et al., 1998).

The use of velvetbean is not a recent innovation. In the U.S., before inorganic nitrogen fertilizer became readily available around 1940, corn systems were also dependent on green manures and living mulches. In 1921, 2.7 million acres of corn in the southern U.S. were interplanted with velvetbean as a fertilizer and feed crop. Maize yield increases of 60 to 80% following velvetbean use were consistently reported (Buckles et al., 1998). Velvetbeans were superior to cowpeas or soybeans for increasing yields of corn, wheat, sorghum, cotton, and oats. Special mechanized equipment was developed to allow interplanting of the two crops (Thurston, 1997). By the 1940s, the acreage in velvetbean dropped due to the widespread availability of nitrogen fertilizers and the replacement of velvetbeans with soybeans for animal feed. By 1965, velvetbean had disappeared from the U.S. agricultural census (Buckles et al., 1998).

Many kinds of cover crops, green manures, and smother crops (for weed control) are in use as living and dead mulches by temperate zone organic farmers today. Twenty-six different species of legumes and grasses are detailed for these uses in mechanized, row crop systems in a current organic farming manual (Zimmer, 2000). In the upper Midwest, kura clover (*Trifolium ambiguum*) is currently being examined for its potential when intercropped with corn as a living mulch (Zemenchik et al., in press). In the mid-Atlantic states, hairy vetch (*Vicia villosa*) has been identified as one of the most productive mulches when grown during the winter months. It produces 2.7 to 4.4 t/ha of dry matter when slashed with a high speed flail mower. This biomass contains sufficient N (90 to 180 kg/ha) so that no commercial fertilizers are necessary for a commercial crop of tomatoes (40 t/ha) in the following summer (Abdul-baki and Teasdale, 1993). When hairy vetch is planted as a winter mulch, the yields of succeeding crops of snap beans and sweet corn are also comparable to those of conventional systems (Abdul-baki and Teasdale, 1995).

## 5.5 SUSTAINABILITY OF MULCH SYSTEMS: CONCLUSIONS

Mulch is an important component of many of the world's traditional systems and modern temperate organic systems, but has been overlooked by commercial temperate zone agriculture. In the tropics it has been confused with slash and burn, which has received more attention because it is more dramatic. The profound effects of mulch on soil fertility and plant growth belie the subtle interaction of root and rooting medium, whether in soil or decomposing organic matter. The ability of mulch to protect the soil against the erosive effects of rainfall is also critical to sustainability. In summary, mulches are critical components of sustainable agricultural systems due to their positive effects on nutrient cycling and energy flow. They should become more widely appreciated with the expansion of sustainable agroecosystems.

## ACKNOWLEDGMENTS

The assistance of Darryl Cole, owner of Finca Loma Linda, Dr. Jim Kettler and Dr. Ken Schlather, and numerous skilled workers and volunteers in Costa Rica is very much appreciated. Special thanks to University of California students J. Loda and C. Reese for initiating the nematode and arthropod work with their independent studies. The research was made possible by funding from the Jessie Smith Noyes Foundation, the Organization for Tropical Studies, and the Cornell University Mulch Based Agriculture Program. The thorough editing of Dr. Eric Engles is gratefully acknowledged, as is Dr. Steve Gliessman's encouragement and support during the first years of my work on the slash mulch system.

## REFERENCES

Abawi, G., Root rots, in Schwartz, H.F. and Pastor-Corrales, M.A., Eds., *Bean Production Problems in the Tropics*, CIAT, Cali, Colombia, 1989, 105–157.

Abawi, G.S. and Pastor-Corrales, M.A., Root rot of beans in Latin America and Africa: Diagnosis, research methodologies, and management strategies, CIAT, Cali, Colombia, 1990.

Abawi, G.S. and Thurston, H.D., Effects of organic mulches, soil amendments and cover crops on soilborne plant pathogens and their root diseases: a review, in Thurston, H.D., Smith, M., Abawi, G., and Kearl, S., Eds., *Slash/Mulch: How Farmers Use It and What Researchers Know About It*, Cornell University, Ithaca, NY, 1994, 88–99.

Abdul-baki, A. and Teasdale, J.R., A no-tillage tomato production system using hairy vetch and subterranean clover mulches, *HortScience*, 28, 106–108, 1993.

Abdul-baki, A. and Teasdale, J.R., Establishment and yield of sweet corn and snap beans in a hairy vetch mulch, *Proc. Fourth Nat. Symp. Stand Establishment Hort. Crops.*, 111–118, 1995.

Aponte, A., Perez, A., and Tablante, J., Control de malezas y plagas en tomate con la utilización de residuos de cosecha, *Divulga Fondo Nacional de Investigaciones Agropecuarias*, 9(41), 10–5, 1992.

Araya, R. and Gonzales, W., *El Frijol Bajo el Sistema Tapado en Costa Rica.* CIAT, Cali, Colombia, 1986.

Arias, F. and Amador, M., Frijol tapado un sistema ventajoso para el pequeño productor, *Avances de Investigación*, 4, 11–16, 1990.

Arias, J., Chacón, A., and Meléndez, G.R., Valoración financiera y de riesgo de los sistemas de siembra de frijol tapado y espequeado, in Meléndez, G., Vernooy, R., and Briceño, J., Eds., *El Frijol Tapado en Costa Rica: Fortalezas, Opciones y Desafíos*, Asociacíon Costarricense de Ciencia de Suelo, San José, 1999, 129–156.

Asiegbu, J.E., Response of tomato and eggplant to mulching and nitrogen fertilization under tropical conditions, *Scientia Horticulturae*, 46(1–2), 33–41, 1991.

Baker, R. and Martinson, C.A., Epidemiology of diseases caused by *Rhizoctonia solani*, in J.R. Parmeter, Jr (Ed.), *Rhizoctonia solani: Biology and Pathology*, University of California Press, Berkeley, CA, 1970, 172–188.

Bellows, B., The sustainability of steep land bean farming (*Phaseolus vulgaris*) in Costa Rica: an agronomic and socioeconomic assessment, Ph.D. dissertation, University of Florida, Gainesville, 1992.

Bohringer, A., The potential of alley cropping as a labor efficient management option to control weeds: a hypothetical case, *Tropenlandwirt*, 92(1), 3–12, 1991.

Buckles, D., Velvetbean: a "new" plant with a history. CIMMYT (International Maize and Wheat Improvement Center), internal document, Mexico, 21, 1994.

Buckles, D., Triomphe B., and Sain, G., *Cover Crops in Hillside Agriculture*, International Maize and Wheat Improvement Center, Mexico D.F., Mexico, 1998.

Budelman, A., The performance of the leaf mulches of *Leucaena leucocephala*, *Flemingia macrophylla* and *Gliricidia sepium* in weed control, *Agroforestry Syst.*, 6(2), 137–45, 1988.

Bunch, R., The potential of slash/mulch for relieving poverty and environmental degradation, in Thurston, H. D., Smith, M., Abawi, G., and Kearl, S., Eds., *Slash/Mulch: How Farmers Use it and What Researchers Know About It*, Cornell University, Ithaca, NY, 1994, 5–9.

Bunch, R., Principles of agriculture for the humid tropics: an odyssey of discovery, *ILEIA Newsletter*, 2(3), 18–19, 1995.

Camelo, L.G. de L., Piccolo, G.A., Rosell, R.A., and Heredia, O.S., Phosphorus sustainability in tropical ecosystems, *Scie. Total Environ.*, 192, 75–82, 1996.

Campbell, B., Frost, P., Kirchmann, H., and Swift, M., A survey of soil fertility management in small-scale farming systems in eastern Zimbabwe, *J. Sustainable Agric.*, 11(2/3), 19–39, 1998.

Cardona, C., Insects and other invertebrate bean pests in Latin America, in Schwartz, H.F. and Pastor-Corrales, M.A., Eds., *Bean Production Problems in the Tropics*, 2nd ed., CIAT, Cali, Colombia, 1989, 505–570.

Castro, M.T., Pitre, H.N., Meckenstock, D.H., and Gomez, F., Influence of slash and burn and slash and mulch practices on insect pests in intercropped sorghum and maize in southern Honduras, *Ceiba*, 39(2), 175–81, 1998.

Coleman, D. and Crossley, Jr., D.A., *Fundamentals of Soil Ecology*, Academic Press, New York, 1996.

Cook, S., Dyhrman, S., and Ingram, K., The abundance of detritivores in *frijol tapado vs. espequeado plots*, Dartmouth Tropical Biology Course Book, 15–23, 1993.

CONIF (Corporación Nacional de Investigación y Fomento Forestal), Recuperación de tierras invadidas por el *Imperata contracta* (H.B.K.) Hitchc. a partir de la incorporación de la leguminosa, *Mucuna deeringiana* (Bort.) Small en Urabá, Bogotá, 1987.

Daisley, L.E.A., Chong, S.K., Olsen, F.J., Singh, L., and George, C., Effects of surface-applied grass mulch on soil water content and yields of cowpea and eggplant in Antigua, *Tropical Agric.*, 65(4), 300–4, 1988.

El-Swaify, S.A., Soil erosion and conservation in the humid tropics, in Pimentel, D., Ed., *World Soil Erosion and Conservation*, IUCN, Cambridge University Press, Cambridge, 1993, 233–255.

Ewel, J., Berish, C., Brown, B., Price, N., and Raich, J., Slash and burn impacts on a Costa Rican wet forest site, *Ecology*, 62, 816–829, 1981.

Fitt, B.D.L. and McCartney, H.A., Spore dispersal in splash droplets, in Ayres, P. G. and Boddy, L., Eds., *Water, Fungi and Plants*, Cambridge University Press, Cambridge, 1986, 87–104.

Fluck, R.C., *Energy: The Hidden Input*, Southern Regional Workshop Evaluating Sustainability, Sustainable Agriculture Research and Education, USDA and ACE, Gainesville, FL, 1995.

Galindo, J.J., Abawi, G.S., Thurston, H.D., and Galvez, G., Effect of mulching on web blight of beans in Costa Rica, *Phytopathology*, 73, 610–615, 1983.

Garrity, D.P., Sustainable land-use systems for sloping uplands in Southeast Asia, ASA Special Publication 56, *Amer. Soc. Agron.*, Madison, WI, 41–66, 1993.

Gliessman, S. R., *Agroecology: Ecological Processes in Sustainable Agriculture*, Sleeping Bear Press, Chelsea, MI, 1998.

Griggs, T., Soil conservation starts at the grass roots, Partners in Research for Development, 8, 16–21, 1995.

Haggar, J., Nitrogen and Phosphorous dynamics of systems integrating trees and annual crops in the tropics, Ph.D. dissertation, Cambridge University, UK, 1990.

Hallman, G. and Andrews, K.L., *Manejo Integrado de Plagas Insectiles en la Agricultura*, in Andrews, K.L. and Quezada, J.R., Eds., Departamento de Protección Vegetal, Escuela Agrícola Panamericana, El Zamorano, Honduras, 523–546, 1989.

Hamilton, L.S., Does "deforestation" always result in serious soil erosion? *International Symposium on Management of Rain Forest in Asia*, Oslo, 102–120, March 23–26, 1994.

Henrot, J. and Brussaard, L., Determinants of *Flemingia congesta* and *Dactyladenia barteri* mulch decomposition in alley cropping systems in the humid tropics, *Plant and Soil*, 191(1), 101–107, 1997.

IITA, *IITA Annual Report 1993*, IITA, Ibadan, Nigeria, 1993.

Ikuenobe, C.E., Chokor, J.U., and Isenmila, A.E., Influence of method of land preparation on weed regeneration in cowpea (*Vigna unguiculata* L. Walp.), *Soil and Tillage Res.*, 31(4), 375–83, 1994.

Jalota, S.K. and Prihar, S.S., *Reducing Soil Water Evaporation with Tillage and Straw Mulching*, Iowa State University Press, Ames, IA, 1998.

Jordan, C., *Nutrient Cycling in Tropical Forest Ecosystems: Principles and their Application in Management and Conservation*, John Wiley & Sons, New York, 1985.

Kettler, J., Weeds in the traditional slash/mulch practice of *frijol tapado*: indigenous characterization and ecological implications, *Weed Res.*, 36, 385–393, 1996.

Kettler, J., Fallow enrichment of a traditional slash/mulch system in southern Costa Rica: comparisons of biomass production and crop yield, *Agroforestry Syst.*, 35, 165–176, 1997a.

Kettler, J., A pot study investigating the relationship between tree mulch decomposition and nutrient element availability, *Communications in Soil Science and Plant Analysis*, 28(15&16), 1269–1284, 1997b.

Kumar, D.P., Subbarayappa, A., Hiremath, I.G., Khan, M.M., and Sadashiviah, M.G., 1989. Use of coconut coir-pith: a biowaste as soil mulch in cashew plantations, *Cashew*, 3(3), 23–4.

Kwapata, M.B., Response of contrasting tomato cultivars to depth of applied mulch and irrigation frequency under hot, dry tropical conditions, *Trop. Agric.*, 68(3), 301–303, 1991.

Lakshmanan, R., Punnoose, K.I., Matthew, M., Mani, J., and Pothen, J., Polythene mulching in rubber seedling nursery, *Indian J. Natural Rubber Res.*, 8(1), 13–20, 1995.

Lal, R., Role of mulching techniques in tropical soil and water management, *IITA Tech. Bull.*, 1, 1975.

Lal, R., Soil management systems and erosion control, in Greenland, D.J. and Lal, R., Eds., *Soil Conservation and Management in the Humid Tropics*, John Wiley & Sons, New York, 93–97, 1977.

Lal, R., *Tropical Ecology and Physical Edaphology*, John Wiley & Sons, New York, 1987.

Lal, R., *Soil Erosion in the Tropics. Principles and Management*, McGraw-Hill, New York, 1990.

Lal, R., Erosion control on sloping land with conservation tillage erosion, III Reunión Biennial de la Red Latinoamericana de Labranza Conservacionista, San José, Costa Rica, 82–91, 1996.

Manning, W.J., Crossan, D.F., and Morton, D.J., Effects of planting depth and asphalt mulch on Rhizoctonia root and hypocotyl rot of snapbean, *Plant Dis. Rep.*, 51(3), 158–160, 1967.

Masefield, G.B., The nodulation of annual leguminous crops in Malaya, *Emp. J. Exp. Agric.*, 25, 139, 1957.

Mata, R., Uribe, L., Gadea, A., and Briceño, J., Suelos característicos de tapaderos, in Meléndez, G., Vernooy, R., and Briceño, J., Eds., *El Frijol Tapado en Costa Rica: fortalezas, opciones y desafíos*, ACCS, University of Costa Rica: San José, Costa Rica, 41–78, 1999.

McGlynn, T., Rosemeyer, M.E., and Reese, C., Soil arthropods in unmulched, slash mulched, and tree-mulched bean agroecosystems in Costa Rica, in preparation.

Mele, P. van, Dekens, E., and Gunathilake, H.A.J., Effect of coir dust mulching on weed incidence in a pineapple intercrop under coconut in Sri Lanka, *Mededelingen Faculteit Landbouwkundige en Toegepaste Biologische Wetenschappen*, 61(3b), 1175–1179, 1996.

Meléndez, G. and Szott, L., El barbecho y el funcionamiento biofísico del frijol tapado, in Meléndez, G., Vernooy, R., and Briceño, J., Eds., *El Frijol Tapado en Costa Rica: fortalezas, opciones y desafíos*, ACCS, University of Costa Rica: San José, Costa Rica, 159–184, 1999.

Meléndez, G., Vernooy, R., and Briceño, J., *El Frijol Tapado en Costa Rica: fortalezas, opciones y desafíos*, ACCS, University of Costa Rica: San José, Costa Rica, 1999.

Muimba-Kankolongo, A., Simba, L., Singh, T.P., and Muyolo, G., Outbreak of an unusual stem tip dieback of cassava (*Manihot esculenta* Crantz) in western Zaire, *Agric. Ecosystems Environ.*, 25(2–3), 151–64, 1989.

Neuenschwander, P., Biological control of the cassava mealybug, *Phenacoccus manihoti* (Hom., Pseudococcidae) by *Epidinocarsis lopezi* (Hym., Encyrtidae) in West Africa, as influenced by climate and soil, *Agric. Ecosystems Environ.*, 32(1–2), 39–55, 1990.

Oldeman, L.R., Hakkeling, R.T.A., and Sombroek, W.G., World map of the status of human-induced soil degradation: an explanatory note. International Soil Reference and Information Center, Wageningen, The Netherlands, 1991.

Onsondo, J.M., Management of cabbage black rot (*Xanthomonas campestris*) in Kenya, *Trop. Pest Manag.*, 33(1), 5–6, 1987.

Palmer, A.F.E., Nitrogen and phosphorus responses and yield trends for continuous maize grown under conservation tillage in the lowland tropics, Kang, B.T. and Vander Heide, J., Eds., *Nitrogen management in farming systems in humid and subhumid tropics*, International Institute of Tropical Agriculture, Haren, The Netherlands, 235–245, 1985.

Pankurst, C.E., Biological indicators of soil health and sustainable productivity, in Greenland, D.J., and Szabolcs, I., Eds., *Biodiversity: Measurement and Estimation*, Chapman and Hall, London, 65–73, 1994.

Pastor-Corrales, M.A. and Tu, J.C., Anthracnose, in Schwartz, H.F., and Pastor-Corrales, M.A., Eds., *Bean Production Problems in the Tropics*, CIAT, Cali, Colombia, 77–104, 1989.

Pimentel, D., Overview, in Pimentel, D., Ed., *World Soil Erosion and Conservation*, Cambridge University Press, Cambridge, 1–6, 1993.

Porras, V.H. and Sanchez, J.A., Efecto de coberturas en la base del árbol de cacao en la diseminación de *Phytophthora*, *Turrialba*, 41(4), 589–97, 1991.

Posner, J.L., Cropping systems and soil conservation in the hill areas of tropical America include soil loss and erosion, *Turrialba*, Instituto Interamericano de Ciencias Agricolas, San José, Costa Rica, 32(3), 287–299, 1982.

Ramos, M.L.G. and Boddey, R.M., Yield and nodulation of *Phaseolus vulgarus* and the competitivity of an introduced *Rhizobium* strain: effects of lime, mulch and repeated cropping, *Soil Biol. Biochem.*, 19(2), 171–177, 1987.

Ribier, V., Barahona, M., Damais, G., Mora, H., Munguía, S., and Saénz, C., Estudio sistemático de la realidad agraria de una microregión de Costa Rica: Nicoya-Hojancha IV. Sistema de granos básicos, Universidad Nacional, Heredia, Costa Rica, 44–51, 1988.

Rosemeyer, M.E., The *frijol tapado* experiment and the extent of leaf spot disease, Organization for Tropical Studies Course Book, OTS, San José, Costa Rica, 406–409, 1985.

Rosemeyer, M.E., The effects of different management strategies on the tripartite symbiosis of bean (*Phaseolus vulgarus L*) with *Rhizobium* and vesicular-arbuscular mycorrhyzal fungi in two agroecosystems in Costa Rica, Ph.D. dissertation, University of California, Santa Cruz, 1990.

Rosemeyer, M.E., Yield, nodulation and mycorrhizal establishment in slash/mulch vs. row-cropped beans, in Thurston, D., Smith, M., Abawi, G., and Kearl, S., Eds., *Slash/Mulch: How Farmers Use It and What Researchers Know About It*, CIIFAD and CATIE, Cornell University, Ithaca, NY, 1994, 169–178.

Rosemeyer, M.E., El mantillo vivo en un sistema orgánico de frijol tapado, in Garcia, J. and Najera, J., Eds., *Las Memorias del Simposio Centroamericano Sobre Agricultura Orgánica*, Universidad Estatal a la Distancia, San José, Costa Rica, March 6–11, 1995, 141–162.

Rosemeyer, M.E., Eficiencia de aplicaciones de fósforo en los sistemas frijol tapado y espequeado a traves de tres años, in *Memoria del Taller Internacional Sobre Bajo Fósforo en el Cultivo de Frijol*, Oficina de Publicaciones de la Universidad de Costa Rica, San José, 1996, 157–163.

Rosemeyer, M.E. and Gliessman, S.R., Modifying traditional and high-input agroecosystems for optimization of microbial symbiosis: a case study of dry beans in Costa Rica, *Agric. Ecosystems Environ.*, 40, 61–70, 1992.

Rosemeyer, M.E., Schlather, K., and Kettler, J., The *frijol tapado* agroecosystem: the survival and contribution of a managed fallow system to modern Costa Rican agriculture, in Hatch, U. and Swisher, M., Eds., *Managed Ecosystems: The Mesoamerican Experience*, Oxford University Press, 1999a.

Rosemeyer, M.E., MacGuidwin, A., Hogg, D., Young, D., McGlynn, T., Reese, C., Loda, J., and Posner, J., *Biodiversity in low- and high-input agroecosystems: a tropical and temperate comparison*, Abstract presented to the Ecological Society of America, Annual Meeting, 1999b.

Rosemeyer, M.E., Viaene, N., Swartz, H., and Kettler, J., The effect of slash/mulch and alleycropping bean production systems on soil microbiota in the tropics, *Appl. Soil Ecol.*, in press.

Rosemeyer, M.E. and Kettler, J., Changes in secondary and weed vegetation in unmulched, slash mulched and alleycropped plots, Costa Rica, in preparation.

Rosemeyer, M.E. and Woike, E., Rooting and nodulation dynamics in the slash mulch system of beans enriched with alleycropped nitrogen-fixing trees, in preparation.

Ruissen, M.A., The development and significance of vesicular-arbuscular mycorrhizas as influenced by agricultural practices, in Sylvia, D.M., Hung, L.L., and Graham, J.H., Eds., *Mycorrhizae in the Next Decade*, IFAS Publication Office, Gainesville, FL, 1982, 57.

Rumoroso, M., and Torres, R., Analysis comparativo de costos de producción en frijol, para los métodos de siembra: tapado espequeado y semi-mechanizada, in Meléndez, G., Vernoy, R., and Briceño, J., Eds., *El Frijol Tapado en Costa Rica: fortalezas, opciones y desafíos*, ACCS, San José, Costa Rica, 1999, 115–128.

Schlather, K.J., Eficiencia de fósforo en frijol tapado vs. frijol espequeado, in Garcia, J. and Najera, J., Eds., *Las Memorias del Simposio Centroamericano Sobre Agricultura Orgánica*, Universidad Estatal a la Distancia, San José, Costa Rica, March 6–11, 1995, 163–176.

Schlather, K.J., The dynamics and cycling of phosphorous in mulched and unmulched bean production systems indigenous to the humid tropics of Central America, Ph.D. dissertation, Dept. of Soil, Crop and Atmospheric Sciences, Cornell University, Ithaca, NY, 1998.

Schlather, K.J. and Rosemeyer, M.E., Phosphorus utilization of the yield in the slash mulch and unmulched system in Costa Rica, in preparation.

Singh, A., Singh, K., and Singh, D.V., The successful use of intercropping for weed management in medicinal yam (*Dioscorea floribunda* Mart and Gal), *Trop. Pest Manage.*, 32(2), 105–107, 1986.

Singh, A., Singh, K., and Singh, D.V., Suitability of organic mulch (distillation waste) and herbicides for weed management in perennial aromatic grasses, *Trop. Pest Manage.*, 37(2), 162–5, 1991.

Solórzano, R., de Camino, R., Woodward, R., Tosi, J., Watson, V., Vásquez, A., Villalobos, C., and Jiménez, J., *Accounts Overdue: Natural Resource Depreciation in Costa Rica*, Tropical Science Center, San José, Costa Rica and World Resources Institute, Washington, D.C., 1991.

Sprent, J.I. and Minchin, F.R., Environmental effects on the physiology of nodulation and nitrogen fixation, in Jones, D.G. and Dawes, D.R., Eds., *Temperate Legumes: Physiology, Genetics and Nodulation*, Pitman Advanced Publications, Boston, 1983, 267–317.

Stark, N.M. and Jordan, C.F., Nutrient retention by the root mat of an Amazonian rain forest, *Ecology*, 59, 434–437, 1978.

Talekar, N.S., Influence of cultural pest management techniques on the infestation of sweet potato weevil, *Insect Sci. Appl.*, 8(4–6), 809–14, 1987.

Thurston, D., *Sustainable Practice for Plant Disease Management in Traditional Farming Systems*, Westview Press, Boulder, CO, 1992.

Thurston, D., *Slash/Mulch Systems: Sustainable Methods for Tropical Agriculture*, Westview Press, Boulder, CO, 1997.

Tian, G., Brussaard, L., and Kang, B.T., Biological effects of plant residues with contrasting chemical compostions under humid tropical conditions: Effects on soil fauna, *Soil Biol. Biochem.*, 25(6), 731–737, 1993.

Triomphe, B., Seasonal nitrogen dynamics and long–term changes in soil properties under the mucuna/maize cropping system on the hillsides of Northern Honduras, Ph.D. thesis, Cornell University, Ithaca, NY, 1996.

Unger, P.W., Straw-mulch effects on soil temperature and sorghum germination and growth, *Agron. J.*, 70, 858–864, 1978.

Wellman, F.L., *Coffee: Botany, Cultivation, and Utilization*, Interscience, New York, 1961.

Wilson-Rummenie, A.C., Radford, B.J., Roberstson, L.N., Simpson, G.B., Bell, K.L., Reduced tillage increases population density of macrofauna in a semiarid environment in Central Queensland, *Environ. Entomol.*, 28(2), 163–172, 1999.

Woike, E., *Rooting patterns of beans grown under the frijol tapado system in Las Vegas de Acosta*, Fullbright Scholarship final report, Institute of International Education, New York, 1997.

Yeates, G.W., Wardle, D.A., and Watson, R.N., Relationships between soil microbial biomass and weed-management strategies in maize and asparagus cropping systems, *Soil. Biol. Biochem.*, 25(7), 869–876, 1993.

Yamoah, C.F., Agboola, A.A., and Mulongoy, K., Decomposition, nitrogen release and weed control by prunings of selected alley-cropping shrubs, *Agroforestry Syst.*, 4(3), 239–246, 1986.

Zemenchik, R.A., Albrecht, K.A., Boerboom, C.M., and Lauer, J.G., Corn production with Kura clover as a living mulch, *Agronomy*, 92, (July/August), in press.

Zimmer, G.F., *The Biological Farmer*, Acres U.S.A., Austin, TX, 2000.

# Section II
# Assessing Sustainability

CHAPTER 6

# Nitrogen and the Sustainable Village

Erle C. Ellis, Rong Gang Li, Lin Zhang Yang, and Xu Cheng

## CONTENTS

### 6.1 INTRODUCTION

Use of synthetic nitrogen fertilizer has increased 15-fold over the past 50 years, helping triple world grain production in support of doubled human populations (Constant and Sheldrick, 1992). While synthetic N has boosted grain production in the past, the high yields of many modern N intensive cropping systems now appear unsustainable (Cassman et al., 1995). Moreover, N fertilizers are increasingly harming both local and global environments (Galloway et al., 1995; Ma, 1997).

Nearly half the people on Earth live in rural villages that depend on subsistence agriculture for food (Marsh and Grossa, 1996). Asia, home of most of these populations, now applies about half the global supply of N fertilizer. This proportion is increasing, along with nitrate pollution of groundwater, N saturation of aquatic and terrestrial ecosystems, and reactive N emissions to the atmosphere that are driving global warming and depleting stratospheric ozone (Galloway et al., 1995).

China's subsistence agriculture population (>800 million) is greater than that of any other nation. China now applies more synthetic N than any other nation, about 25% of the world's supply in 1990 (Constant and Sheldrick, 1992; Galloway et al., 1996), and more than Africa and the Americas combined.

This chapter explores the long-term ecological impact of synthetic N use within a rural village in the Tai Lake Region of China (Xiejia Village, Wujin County, Jiangsu Province; Latitude 31.5°N, Longitude 120.1°E; Ellis and Wang, 1997; Ellis et al., 2000a; Ellis et al., 2000b). The broad and unexpected effects of synthetic N within this densely populated village landscape offer valuable lessons for developing agroecosystems that can maintain food security in subsistence agriculture regions with less harm to local and global ecosystems.

## 6.2 NITROGEN IN VILLAGE ECOSYSTEMS

To measure the impacts of synthetic N in village ecosystems, observations are needed across the many different land managers and land types within these systems. Intensive subsistence agriculture clusters large numbers of farmers within relatively small areas, generating highly heterogeneous anthropogenic landscapes with ecosystem processes that are controlled as much by social dynamics as by environmental factors. The interplay between management variability and landscape heterogeneity in village ecosystems generates emergent properties that can only be understood by a village scale approach that incorporates diversity across farmers and landscapes. Figure 6.1 illustrates this situation: farming households with differing management styles, represented by shades of gray, may manage similar and/or neighboring land types, while differing types of land are often managed by the same farmers.

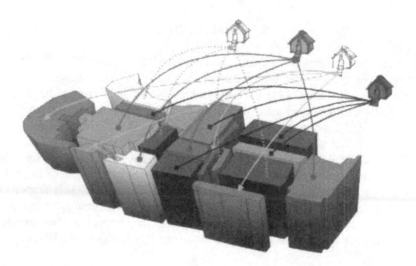

**Figure 6.1**   Concept diagram illustrating the interaction of management diversity and landscape heterogeneity in anthropogenic landscapes.

We will use two examples to demonstrate our village scale approach to assessing and remedying the negative impacts of synthetic N fertilizers in subsistence village ecosystems. We will show that N loading and losses from paddy fields vary greatly between land managers, and that solutions to N overapplication require a full understanding of the basis for this management variability. We will demonstrate that adoption of synthetic N fertilizers has changed soil N sequestration, a key biogeochemical process, across entire village landscapes, even in many areas where fertilizers are not applied.

## 6.3 NITROGEN IN PADDY AGROECOSYSTEMS

Traditional rice/wheat paddy double cropping systems sustained 4 Mg/ha rice yields for centuries in the Tai Lake region, earning the area the title "land of fish and rice" (Ellis and Wang, 1997). In the 1930s, N inputs to these systems in the form of traditional organic amendments were less than 100 kg N/ha/yr, about the same as N removal in grain and straw; there was no significant N runoff or leaching. Synthetic N applications began increasing rapidly in the 1960s, reaching 500 kg N/ha/yr in the late 1980s. By 1994, synthetic N had increased from 0% of paddy fertilizer N to more than 80%, displacing traditional organic inputs (Ellis and Wang, 1997). Over the same period, nitrate pollution and eutrophication became serious regional problems (Ma, 1997).

Relationships between N fertilizer loading, yield, and N losses in rice/wheat paddy fields are illustrated in Figure 6.2, along with 1994 N fertilizer inputs (chemical plus organic) by surveyed farmer households in Xiejia village. If relationships between N loading, loss, and yield were linear, every increment in N loading would boost both yields and losses. However, as Figure 6.2 illustrates, these relationships are nonlinear, and are best described by three phases: a limiting phase, in which yields can be increased without major N losses, an optimal phase, in which yields are maximized, and a saturating phase, in which every increment in N loading intensifies N losses while diminishing yields.

N applications by a surprising number of farmers, about 20%, are in the saturating phase, with yields reduced significantly by N overapplication (Figure 6.2). Although the average N loading for village farmers is within the optimal phase (~480 kg N/ha/yr), about half of village farmers apply more than the average and are contributing to N losses without any possible yield benefit. As a result of the nonlinear relationship between N loading and loss, using the average farmer N loading to estimate village paddy N losses underestimates the true value of such losses by ~5% compared with the average across each farmer's paddy land ~184 vs. 194 kg N/ha/yr; Village households have about the same amount of grain land in China.

To reduce N losses without reducing yields, an extension program might encourage all farmers to limit their N inputs to the current annual average (~480 kg N/ha/yr). According to the simple N loading/loss model of Figure 6.2, this strategy could reduce village paddy N losses by ~19% of their current total without reducing yields. However, the wide variability of household N inputs suggests a solution that might require less effort.

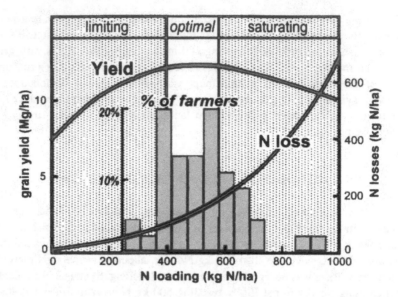

**Figure 6.2**  Annual N loading to paddy fields, yields of rice and wheat (unprocessed), and N loss (including denitrification, leaching, and runoff), superimposed on a histogram of N loading by a sample of 50 farmer households surveyed in Xiejia Village, 1994 (Ellis et al., 2000b).

When household N inputs are divided into four groups based upon input levels, as shown in Figure 6.3, it is evident that the group applying the most N uses much greater amounts than the others. By reducing the inputs of this highest group, comprising only 25% of village farmers, N losses would be reduced by ~16%, nearly the same as if all farmers reduced their inputs to the average.

Farmers in the highest input group also have proportionately higher organic N inputs than other farmers (Figure 6.3; R2 = 0.46). Village animal managers tend to overapply both organic and chemical nutrients as "insurance" because it is difficult to estimate manure nutrient content and because manure storage is limited. This same effect has been observed in North America (Nowak et al., 1998).

To generalize, whenever *farmer types* with consistently higher nutrient inputs (such as *animal managers*) can be identified, programs to lower nutrient losses can maximize their success by targeting these farmers over the bulk farmer population. This strategy, however, depends on understanding the full range of fertilizer management by farmers; using averaged data both obscures the biogeochemistry of N losses and eliminates the management information needed to eliminate these losses.

## 6.4 NITROGEN SEQUESTRATION IN VILLAGE SOILS

The majority of N in most ecosystems is stored in soil organic compounds (Stevenson, 1986). As a result, relatively small changes in soil N storage can transform landscapes from sources to sinks of N, with potentially global implications (Simpson et al., 1977). Soil N storage varies considerably across landscapes, influenced by

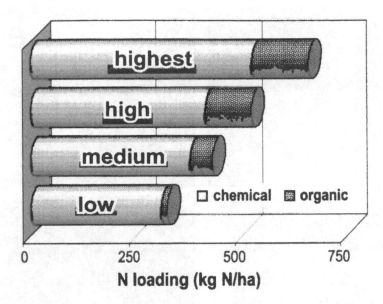

**Figure 6.3** Annual N loading to paddy fields by 50 households in Xiejia village, 1994, separated into quartiles based on their amount of N loading and averaged within quartiles. Chemical inputs are urea, ammonium bicarbonate, and compound fertilizers; organic inputs are human and animal manures.

such factors as terrain (erosion and runoff), management (fertilizer, harvest, burning), and hydrology (dry areas leach more nitrate, moist areas have greater ammonia volatilization and denitrification) (Stevenson, 1982). See Figure 6.4.

To measure long-term changes in soil N storage, subsistence village landscapes must first be stratified into relatively homogeneous landscape components for

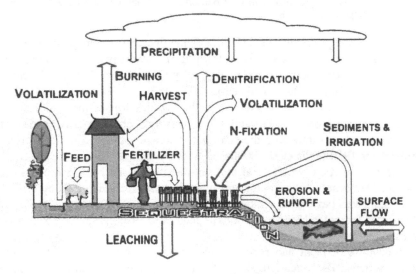

**Figure 6.4** Biogeochemistry of nitrogen across village landscapes.

**Figure 6.5**   Fine scale heterogeneity in village landscapes. (A) Village landscapes near Tai Lake, Wuxi, China 1924, illustrating rice paddy fields, paddy bunds, paths, houses, upland plots, fish ponds, village canals and individual mature trees; pond edges are lined with mulberry trees (Buck, 1937). (B) Canals, houses and upland plots in Xiejia village, 1994; note fine scale landscape management, including individual evergreen trees. (C) Harvesting soybeans from field borders; soybeans and broad-beans are grown only in field borders. (D) Canal nearly filled with sediment in Xiejia village, 1995.

sampling and analysis. Figure 6.5 illustrates the fine scale heterogeneity of village landscapes. Figure 6.6 presents the anthropogenic landscape classification system used to stratify heterogeneity into ecologically homogenous "ecotope" landscape components in Xiejia village (Ellis et al., 2000b). By measuring soil and sediment N storage in the top 40 cm of soil and in the low density sediments (<1.3 g/cm$^3$) of each village ecotope in 1930 and 1994, long-term changes in N sequestration were calculated by subtracting the ecotope N storage estimates for 1994 from the estimates for 1930 (Ellis et al., 2000a).

N storage in Xiejia village soil and sediment increased by ~25% overall from 1930 to 1994 because N concentrations in agricultural soils increased by ~20% and N-rich sediments have filled village canals (Ellis et al., 2000a) Figure 6.5D shows such a canal. Increased N concentrations in paddy and other agricultural soils account for about half of the total village soil N storage increase; they are best explained by the stimulation of plant and soil biomass production by synthetic N subsidy of agroecosystems. The remaining half of the N buildup was caused by sediment accumulation since the end of communal agriculture in 1982, when inexpensive

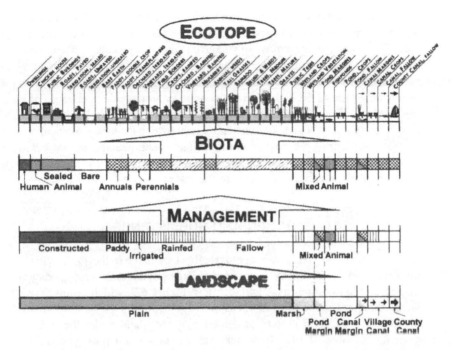

**Figure 6.6** Anthropogenic landscape classification hierarchy for all ecotopes in Xiejia village (Ellis et al., 2000b).

synthetic N replaced the traditional labor intensive practice of harvesting canal sediments for fertilizer. At current rates, sediments will completely fill most village canals within 25 years, increasing flood risk and impeding irrigation and transport. This unanticipated environmental impact is an indirect result of the transition from traditional organic fertilizers to synthetic N.

In the traditional system, conflicts arose over rights to use nutrient rich sediments for fertilizer. Now, there are clashes over who is responsible for clearing irrigation canals when water fails to reach irrigation pumps as it did in the regional drought of 1994. What can be done with nearly 30 tons of sediment entering canals, marshes, and ponds every year for every hectare of village land? Some efforts have been made to mechanize the use of sediments for fertilizer, but farmers have little incentive to use these methods when inexpensive fertilizers are such convenient, labor saving substitutes. It is likely that money will have to be spent to clear sediments in village irrigation canals — an unforeseen cost of synthetic N that will have to be accounted for eventually.

Synthetic N has also displaced another traditional fertilizer — nightsoil (human manure). In the past, most nightsoil was applied primarily to paddy land, at rates rarely exceeding 40 kg N/ha/yr. Now, to save labor, most nightsoil is applied to small upland plots near houses at rates often exceeding 200 kg N/ha/yr, transforming nightsoil from a valued fertilizer into an excess nutrient input in the drier areas of the village most susceptible to nitrate leaching. It is likely that manure management systems will need to be developed to reduce the labor requirements of composting and spreading these manures over larger areas, most likely at some cost to farmers or the state.

Inexpensive synthetic N, combined with the high demand for rural labor in the Tai Lake region, is driving major changes in N biogeochemistry across subsistence village landscapes. These changes are evident in village canals and wetlands, and in the upland plots that are still fertilized using only manures. To assess the long-term ecological impacts of synthetic N in subsistence agriculture regions, biogeochemical changes must be monitored across the highly heterogeneous anthropogenic landscapes of rural village ecosystems. The economic impact of the displacement of traditional fertilizer management by synthetic N must be considered in assessing the agroecological impact of this input. Considering the cost of sediment and waste management reveals that traditional fertilizer systems provided "agroecosystem services" that must now be replaced, most likely at considerable cost to the state.

## 6.5 NITROGEN AND SUBSISTENCE

Though traditional grain yields were ecologically sustainable in the Tai Lake region, they are insufficient to feed today's doubled village populations. This predicament is illustrated by the question mark and arrow in Figure 6.7, which points from the 1985 population per hectare of paddy land in Wujin county to the number of people who could be fed by grain protein produced by paddy land under the traditional management conditions of 1930 since subsistence populations must generally pro-duce at least twice their minimum food requirements to attain food security (Luyten

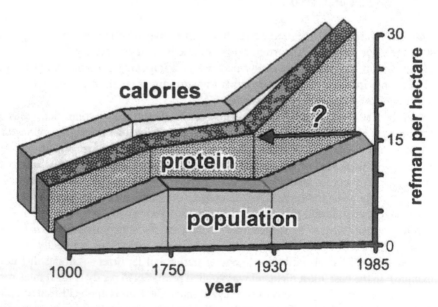

**Figure 6.7**  Human nutritional carrying capacity of rice/wheat systems, 1000 A.D. to 1985. Chinese standard "reference man" (refman) units are used to express population per hectare of paddy land (refman per hectare); these units standardize popula-tions by the size of their nutritional requirements. "Protein" and "calories" describe the number of people who could be fed by the grain protein or calories produced per hectare of paddy land at the times indicated (Ellis and Wang, 1997).

et al., 1997), it is clear from Figure 6.7 that current populations would have inadequate food under traditional production conditions.

Paddy land per capita dropped from 0.11 ha in 1930 to 0.05 ha in 1994 and is still declining. Although they require more than 5 times as much N overall, modern varieties and synthetic N now appear essential in producing the ~12 Mg/ha of rice and wheat that feeds current village populations. The availability of nitrogen limited traditional rice yields, even with the best traditional management using legume green manures, sediment composts, and careful husbanding of animal and human manures. Nitrogen removal in current grain and straw production is about 230 kg N/ha, more than twice the total N loading from traditional fertilizer inputs to paddy land.

These data confirm the basic facts: N is most often the limiting nutrient in ecosystems. Sustaining high grain yields is critical for village food security. To maintain food security in subsistence agricultural regions, N management must sustain high yields without causing excessive environmental damage. Evidence from China's Tai Lake region reveals that synthetic N management can be improved to lessen its negative impacts. There is potential for revitalizing the use of traditional N inputs, such as sediments and nightsoil in order to substitute for much of the N now supplied by synthetic fertilizers; this will require both labor saving technology and political intervention.

## 6.6 CONCLUSIONS

A village scale approach to measuring and mediating the impacts of synthetic N is essential in securing the long-term sustainability of subsistence agriculture. In contrast with regional analyses based on data from the county level and above, village scale analysis can identify both the sources of environmental problems and the pathways toward solving these problems. In the densely populated agricultural landscapes that cover as much as $8 \times 10^6$ km$^2$ of the earth's surface, these methods are necessary both for assessing long-term biogeochemical change and in forming policies that can remedy the negative impacts of these changes. Similar methods for anthropogenic landscape classification and manager level analysis should prove useful in other densely populated anthropogenic landscapes as well, such as those of urban and periurban areas. By monitoring the long-term impacts of synthetic N and other industrial inputs across entire village ecosystems, solutions can be developed that sustain both agricultural productivity and environmental quality for the populations of subsistence agricultural regions.

## ACKNOWLEDGMENTS

We thank the leaders of Xiejia and Zhangqing villages, Mashan and Xueyan townships, Wuxi municipality, Wujin county and especially the Jiangsu Department of Agriculture and Forestry for making this work possible. This material is based upon work supported by the National Science Foundation under grant DEB-9303261 awarded in 1993, and grant INT-9523944 awarded in 1995.

# REFERENCES

Buck, J.L., *Land Utilization in China: Atlas*, Commercial Press, Shanghai, 1937.

Cassman, K.G., De Datta, S.K., Olk, D.C., Alcantara, J., Samson, M., Descalsota, J., and Dizon, M., Yield decline and the nitrogen economy of long-term experiments on continuous, irrigated rice systems in the tropics, in Lal, R., and Stewart, B.A., Eds., *Soil Management: Experimental Basis for Sustainability and Environmental Quality*, CRC Press, Boca Raton, FL, 181–222, 1995.

Constant, K.M. and Sheldrick, W.F., *World Nitrogen Survey*, vol. 174, World Bank, Washington, D.C., 1992.

Ellis, E.C., Li, R.G., Yang, L.Z., and Cheng, X., Changes in village-scale nitrogen storage in China's Tai Lake Region, *Ecological Applications* 10(4), 1074–1089, 2000a.

Ellis, E.C., Li, R.G., Yang, L.Z., and Cheng, X., Long-term change in village-scale ecosystems in China using landscape and statistical methods, *Ecological Applications*, 10(4), 1057–1073, 2000b.

Ellis, E.C. and Wang, S.M., Sustainable traditional agriculture in the Tai Lake Region of China, *Agric. Ecosystems Environ.*, 61(2–3), 177–193, 1997.

Galloway, J.N., Dianwu, Z., Thomson, V.E., and Chang, L.H., Nitrogen mobilization in the United States of America and the People's Republic of China, *Atmospheric Environ.*, 30(10–11), 1551–1561, 1996.

Galloway, J.N., Schlesinger, W.H., Levy, H., II, Michaels, A., and Schnoor, J.L., Nitrogen fixation: anthropogenic enhancement, environmental response, *Global Biogeochemical Cycles*, 9(2), 235–252, 1995.

Luyten, J.C., Shi, Q.H., and Penning De Vries, F.W.T., The limits of consumption and production of food in China in 2030, in Teng, P.S., Kropff, M.J., ten Berge, H.F.M., Dent, J.B., Lansigan, F.P., and van Laar, H.H., Eds., *Applications of Systems Approaches at the Farm and Regional Levels*, 1, Proc. of the 2nd Int. Symp. Syst. Approaches for Agricultural Dev., held at IRRI, Los Banos, Philippines, 6–8 December, Kluwer Academic Publishers, Boston, 1995, 281–293.

Ma, L.S., Nitrogen management and environmental and crop quality, in Zhu, Z.L., Wen, Q.X., and Freney, J.R., Eds., *Nitrogen in Soils of China*, 74, Kluwer Academic Publishers, Dordrecht, The Netherlands, 1997, 303–321.

Marsh, W.M. and Grossa, J.M., Jr., *Environmental Geography: Science, Land Use, and Earth Systems*, John Wiley & Sons, New York, 1996.

Nowak, P., Shepard, R., and Madison, F., Farmers and manure management: a critical analysis, in Hatfield, J.L. and Stewart, B.A., Eds., *Animal Waste Utilization: Effective Use of Manure as a Soil Resource*, Ann Arbor Press, Chelsea, MI, 1998, 1–32.

Simpson, H.J., Broecker, W.S., Garrels, R.M., Gessel, S.P., Holland, H.D., Holser, W.T., Junge, C., Kaplan, I.R., McElroy, M.B., Michaelis, W., Mopper, K., Schidlowski, M., Seiler, W., Steele, J.H., Wofsy, S.C., and Wollast, R.F., Man and the global nitrogen cycle group report, in Stumm, W., Ed., *Global Chemical Cycles and Their Alterations by Man: Report of the Dahlem Workshop on Global Chemical Cycles and Their Alterations by Man*, Berlin Abakon-Verlagsgesellschaft, Berlin, November 15–19, 2, 1977, 253–274.

Stevenson, F.J., Origin and distribution of nitrogen in soil, in Stevenson, F.J., Ed., *Nitrogen in Agricultural Soils*, American Society of Agronomy, Madison, WI, 22, 1982, 1–42.

Stevenson, F.J., Cycles of Soil, *Carbon, Nitrogen, Phosphorus, Sulfur, Micronutrients*, John Wiley & Sons, New York, 1986.

# CHAPTER 7

# Nematode Communities as Ecological Indicators of Agroecosystem Health

Deborah A. Neher

## CONTENTS

## 7.1 INTRODUCTION

In the search for ways to measure the sustainability of agroecosystems, the soil has often been ignored. However, soil supports essential ecosystem functions, such as promoting plant productivity, enhancing water relations, regulating nutrient mineralization, permitting decomposition, and acting as an environmental buffer (Neher,

1999a). The relative health and quality of an agroecosystem's soil, especially its biological and ecological components, should correlate closely with the overall health and sustainability of the system.

Biologically, soil ecosystems support a great diversity of both microfauna (fungi, bacteria, protozoa, and algae) and mesofauna (protozoa, arthropods, and nematodes). The author's research suggests that measures of the community structure of one of these biological components, namely nematodes, may serve as useful indicators of the health of agricultural soils, and therefore have promise as indicators of sustainability.

## 7.2 NEMATODES AND THEIR ROLE IN THE SOIL ECOSYSTEM

Soil nematodes (roundworms) are relatively abundant ($6 \times 10^4$ to $9 \times 10^6$ per m$^2$), small (300 μm to 4 mm) animals with short generation times (days to a few weeks) that allow them to respond to changes in food supply (Wasilewska, 1979; Bongers, 1990). Relative to other soil micro and mesofauna, trophic or functional groups of nematodes can be identified easily, primarily by morphological structures associated with various modes of feeding (Yeates and Coleman, 1982). Nematode species with a buccal stylet (spear like structure) feed on cell contents and juices obtained by piercing the cellular walls of plant roots or fungal mycelium. Other species have no stylets and feed on particulate food such as bacteria and small algae (Vinciguerra, 1979). Agricultural soil communities often have large numbers of bacterial and plant feeding nematodes and smaller numbers of fungal feeding, omnivorous, and predaceous nematodes (Wasilewska, 1979; Hendrix, et al., 1986).

Nematodes are part of a complex soil ecosystem. A small fraction of soil fauna depends upon primary producers, feeding on plant roots and their exudates. The subgroups of these organisms that form parasitic relationships with plants and their roots are the best known of soil organisms because of the damage they cause to agricultural crops. They decrease plant production, disrupt plant nutrient and water transfer, and decrease fruit and tuber quality and size (Yeates and Coleman, 1982; Brussaard et al., 1997). Most soil organisms perform beneficial roles in ecosystem function and are not parasites or pests. For example, most soil bacteria, actinomycetes, fungi, algae, and protozoa are decomposers of organic matter. These microorganisms are involved directly with production of humus, cycling of nutrients and energy flow, elemental fixation, metabolic activity in soil, and the production of complex chemical compounds that cause soil aggregation. Microbial grazing mesofauna (e.g., Collembola, mites, nematodes, and protozoa) affect growth and metabolic activities of microbes and alter the microbial community, thus regulating decomposition rate (Wasilewska et al., 1975; Trofymow and Coleman, 1982; Whitford et al., 1982; Yeates and Coleman, 1982; Seasteadt, 1984) and nutrient mineralization (Seastedt et al., 1988; Sohlenius et al., 1988).

It should be noted that only 10% of soil dwelling species have been identified (Hawksworth and Mound, 1991). Our knowledge of soil organisms has been limited by our inability to extract organisms from soil efficiently and by difficulties in appropriately identifying juvenile stages (Neher, 1999a). Modern soil biologists are involved in identification of new species, determining their food preferences, quantifying interactions among organisms in soil communities, and defining specific functions in ecosystems (de Bruyn, 1997).

Soil dwelling organisms are linked through detrital food webs, which consist of pathways centered on plant roots, bacteria, and fungi (Moore et al., 1988; see Figure 7.1). Protozoa feed primarily within the bacterial pathway. Microarthropods feed primarily within the fungal pathway. Bacterial and fungal pathways unite higher in the food chain at the trophic level of predaceous nematodes and mites (Whitford, 1989). Arthropods and nematodes have the potential to feed upon, or otherwise affect, organisms in all three pathways.

## 7.3 NEMATODE COMMUNITY INDICES

Nematodes have attributes that make them useful as ecological indicators (Freckman, 1988; Neher and Campbell, 1994). Various kinds of perturbations to soils, such as addition of mineral nitrogen fertilizers (Wasilewska, 1989), cultivation (Hendrix et al., 1986), liming (Hyvonen and Persson, 1990), and accumulation of heavy metals (Samoiloff, 1987; Bongers et al., 1991) affect the species richness, trophic structure, and successional status of nematode communities. Because they reflect changes in soil structure and function related to these perturbations, indices of nematode community structure show promise for monitoring the ecological condition of the soil (Bongers, 1990; de Goede, 1993; Ettema and Bongers, 1993; Freckman and Ettema, 1993).

There are many methods of measuring nematode community structure. Through a series of experiments on sampling and experimental design at various spatial scales, the author concluded that maturity (Bongers, 1990) and trophic diversity indices are capable of differentiating among sampling sites better and more efficiently than measures based on populations or ratios of individual trophic groups (Neher et al., 1995). Maturity and trophic diversity indices measure different aspects of soil communities and are complementary when used together. "Maturity" is a measure of successional status and trophic diversity measures food web structure.

### 7.3.1  Maturity Indices

Maturity indices are a measure of the ecological successional status of a soil community. They are based on the principle that different taxa have different sensitivities to stress or disruption of the successional sequence because of differences in their life history characteristics. Because succession can be interrupted at various stages by common agricultural practices, such as cultivation and applications of fertilizer and pesticides (Ferris and Ferris 1974; Wasilewska, 1979), the successional status of a soil community may reflect the history of disturbance.

The maturity index is a weighted mean frequency of taxa assigned weights ranging from 1 to 5, with smaller weights assigned to taxa with relative tolerance to disturbance and larger weights representing taxa more sensitive to disturbance (Bongers, 1990). A maturity index for free living taxa (MI) may be viewed as a

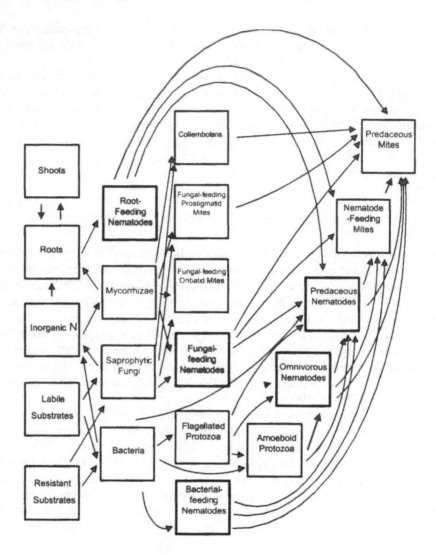

**Figure 7.1**   Soil food web in native North American shortgrass steppe prairie in eastern
Colorado. Arrows indicate potential feeding relationships that were quantified. Five
trophic groups of nematode communities are highlighted. (From Moore, J.C. and
de Ruiter, P.C., Temporal and spatial heterogeneity of trophic interactions within
below ground food webs, in *Agriculture, Ecosystems and Environment*, 34,
391–397, 1991. With permission.)

measure of disturbance, with smaller values indicative of a more disturbed environment and larger values characteristic of a less disturbed environment (Freckman and
Ettema, 1993). A maturity index for plant parasitic taxa (PPI) may or may not
correlate positively with MI (Bongers, 1990; Freckman and Ettema, 1993; Neher
and Campbell, 1994).

Following a disturbance, such as the addition of animal manure to soil, progressive increases in the abundance of nematodes with large maturity index values have

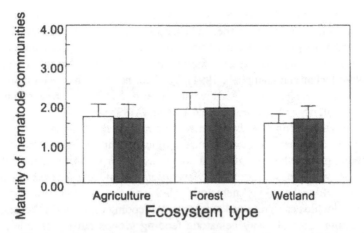

**Figure 7.2** Means and standard errors of nematode community composition measured as successional maturity (ΣMI25) in disturbed (open bars) and undisturbed (solid bars) soils of North Carolina. Disturbed is defined as annually cultivated arable soils, 1- to 3-year-old forests, and functioning wetlands. Undisturbed is defined as 10+-year-old pastures, 30+-year-old forests, and wetlands converted to agricultural production.

been documented, following the initial predominance of nematodes with smaller values (Ettema and Bongers, 1993). Similarly, soil communities in uncultivated or no-till agricultural systems have been considered successionally more mature than those in frequently cultivated agricultural soil (Hendrix et al., 1986; Freckman and Ettema, 1993). Likewise, soils with perennial crops are successionally more mature than soils with annual crops (Freckman and Ettema, 1993; Neher and Campbell, 1994). Interpretation of maturity indices, however, depends on ecosystem type (D. A. Neher, M. E. Barbercheck, and O. Anas, unpublished data). Successional maturity is greater in forest soils than in wetland and agricultural soils (Figure 7.2).

## 7.3.2 Trophic Diversity Indices

Trophic diversity indices describe the relative abundance and evenness (Ludwig and Reynolds, 1988) of the occurrence of five nematode trophic groups (see Figure 7.1). Trophic diversity can be expressed with either a Shannon or a Simpson diversity index (Shannon and Weaver, 1949; Simpson, 1949). In agricultural soils, greater diversity of trophic groups is correlated with an increase in the frequency of occurrence of generally less abundant trophic groups (i.e., fungal feeding, omnivores, and predators) relative to that of generally more abundant trophic groups (i.e., bacterial and plant feeding groups) (Wasilewska, 1979). Due to the typically unequal distribution of trophic groups within nematode communities in agroecosystems (Wasilewska, 1979), the Shannon index — which gives more weight to abundant taxa (Ludwig and Reynolds, 1988) — may be more applicable than the Simpson index of diversity (Neher and Campbell, 1994).

It has been demonstrated that disturbances such as cultivation (Freckman and Ettema, 1993) and addition of manure (Neher and Olson, 1999) decrease trophic diversity. Also, trophic diversity tends to be greater in soils with perennial crops than it is in soils with annual crops (Neher and Campbell, 1994). These apparent differences are attributed mostly to a decline in numbers of omnivorous and predaceous nematodes and increases in numbers of bacterial-feeding nematodes (Wasilewska, 1979; Hendrix et al., 1986; Neher and Campbell, 1994). To date, no study has been published that tracks temporal changes in trophic diversity after a disturbance in as much detail as changes in maturity indices (Ettema and Bongers, 1993).

Appropriate caution must be taken when applying trophic designations to nematode species and genera because recent ecological studies have revealed that feeding habit groupings may be ambiguous in some cases. For example, abundant populations of *Aphelenchoides*, *Tylenchus*, *Tylencholaimus*, and *Ditylenchus* can be classified as "plant/fungal feeding" nematodes (Sohlenius et al., 1977) or some "predaceous" mononchids can grow and reproduce by feeding on bacteria (Yeates, 1987b). Current assignments of many nematode feeding groups have been inferred rather than confirmed by maintenance of nematodes over many generations under biologically defined conditions (Yeates et al., 1993). The problem could be minimized if supplementary studies were conducted to examine critically the feeding preferences of nematode taxa in defined environments.

## 7.4 RELATIONSHIP OF NEMATODE COMMUNITIES TO ECOSYSTEM FUNCTION

Indices of nematode community structure can be considered appropriate and successful indicators of soil quality if they correspond with ecological processes occurring in the soil; that is, nutrient mineralization and decomposition of organic matter. It is too soon to claim that this is the case, because initial experiments have been correlative in nature. Nevertheless, the studies point to a direct relationship between the structure of soil nematode communities and ecological processes.

### 7.4.1  Nutrient Availability

Nematodes have been demonstrated to affect plant productivity by increasing nutrient availability through regulation of mineralization processes. For example, shoot biomass and nitrogen content of plant shoots grown in the presence of protozoans and free living nematodes are greater than those of plants grown without mesofauna (Verhoef and Brussaard, 1990; Yeates and Wardle, 1996). Associations between nematode presence and increased availability of nitrogen were determined in experiments performed in petri dishes (Trofymow and Coleman, 1982), and field studies have confirmed these findings (Neher, 1999b). The basis of this relationship is that grazing on microbes by mesofauna releases and mineralizes nutrients immobilized in microbes, subsequently converting nitrogen from organic to inorganic forms that plants can utilize (Trofymow and Coleman, 1982; Seastedt et al., 1988; Sohlenius et al., 1988).

Soil fauna are responsible for approximately 30% of nitrogen mineralization in agricultural and natural ecosystem soils. Protozoa and bacteria feeding nematodes, the main consumers of bacteria, account for 83% of this nitrogen mineralization (Elliott, et al., 1988), and are estimated to contribute about 8 to 19% of nitrogen mineralization in conventional and integrated farming systems (Beare, 1997). Nematodes contribute directly to nitrogen mineralization by excretion of nitrogenous wastes, mostly as ammonium ions (Anderson et al., 1983; Ingham ct al., 1985; Hunt et al., 1987). In addition to serving as a stimulatory force in net mineralization of nutrients, nematodes also promote nutrient immobilization because their bodies constitute reservoirs of nutrients. When nematodes die nutrients immobilized in their tissues are mineralized and subsequently become available to plants.

Although nematode presence is correlated positively with increased availability of nutrients, the relationship between nematode community structure and nutrient availability is less clear. In agricultural soils, increased nitrogen fertilization, a disturbance to the ecosystem, returns nematode community structure to an earlier successional state, similar to additions of manure (Ettema and Bongers, 1993). Negative correlations occur between successional maturity of nematode communities and concentrations of nitrate and ammonium, two forms of nitrogen available to plants (Neher, 1999b). The causality in this relationship, however, is unclear. In general, the actual mechanisms of soil organisms' impact on soil fertility are not well understood (Giller et al., 1997).

## 7.4.2 Decomposition of Organic Matter

The process of organic matter decomposition is closely linked to nutrient mineralization and immobilization. Cellulose and lignin represent two abundant molecules present in organic matter that must be decomposed. Cellulose is composed of labile compounds and is, therefore, decomposed easily by a wide variety of microorganisms. Lignin is more resistant to decay, and only specialized fungal species can decompose it (Dix and Webster, 1995).

The author conducted an experiment in agricultural systems in North Carolina measuring nematode communities and decomposition of cellulose and lignin as mass loss through time. Positive correlations ($p < 0.05$) were observed between successional maturity of nematode communities and decomposition of cellulose in non-cultivated, perennial agricultural systems (D. A. Neher, M. E. Barbercheck, and O. Anas, unpublished data). This result suggests a direct association between successional maturity of soil communities and ecosystem function. However, this association was decoupled in cultivated soils with annual crops ($p > 0.05$). No correlation was observed between nematode community composition and lignin decomposition.

## 7.5 INDEX CALIBRATION

Ecological indices related to nematode communities do not provide absolute values of condition but require reference to some putatively undisturbed community for interpretation or comparison (de Bruyn, 1997; Neher, 1999b). Use of an undisturbed community for a reference point is unrealistic, because agroecosystems are disturbed intentionally for human purposes.

Two types of agroecosystems, perennial crop systems and organic systems, may serve as bases of comparison when using nematode community indices to measure the health of soils cultivated with annual crops. The argument for using perennial systems as the basis of comparison is based on the claim that farming systems that include soil conservation practices, such as zero or minimum tillage, are more sustainable than those employing conventional practices (de Bruyn, 1997). The argument for using organic systems rests on the claim that organic farms support nematode communities undisturbed by agricultural chemicals.

### 7.5.1  Perennial Crops

Research suggests that forage and pasture agroecosystems may be suitable for use as a reference point in monitoring the ecological condition of soils associated with annual crops. The PPI and the ratio of fungal feeding to bacterial feeding nematodes for annual crops (soybean, corn, wheat) were found to be significantly different from those for perennial crops (alfalfa and tall fescue grass, alone or mixed with legumes). Trophic diversity was similar in the two systems, suggesting no differences in food web structure. Results indicate that soils with perennial crops are more mature successionally than soils with annual crops and that the ratio of fungal to bacterial feeding nematodes may be an important description of the decomposition pathway in detritus food webs (Sohlenius and Sandor, 1987). Data from this study are summarized in Figure 7.3.

When using a perennial system as a base of comparison, the period of time that the system has been undisturbed must be considered (Wasilewska, 1979; 1994). A long-term perennial crop (> 10 years) is related more closely to an undisturbed site than a medium term (2 to 5 years) perennial crop. Crop land that has never been cultivated or has been abandoned for a long period may be the best reference (Freckman and Ettema, 1993). These sites may be difficult to locate. Perennial agriculture fields such as those described above would be practical choices for a large scale monitoring program because they occur frequently in agroecosystems and are widespread geographically (Neher and Campbell, 1994).

### 7.5.2  Organic Production

Organic management minimizes or excludes synthetic chemical fertilizers. It places an emphasis on the recycling of organic wastes and the use of legume crops as green manure to supply nutrients on a schedule matching plant demand (USDA, 1980). In organic farming systems, soil microbes appear to play a more important role in plant nutrient cycling than they do in conventional systems (Allison, 1973). Farm practices that minimize the use of synthetic pesticides or inorganic fertilizers generally result in soils with ecological properties considered "good" or "healthy" (Bolton, 1983; Reaganold et al., 1993).

The author participated in two studies that compared successional maturity and trophic diversity of nematode communities in organic and conventional farming systems. The first experiment compared soils from 5 pairs of organically and conventionally managed soils in the Piedmont region of North Carolina (Neher, 1999b). The

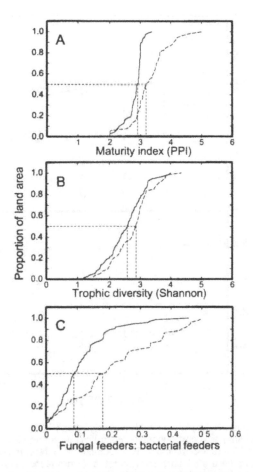

**Figure 7.3** Cumulative distribution function of the (A) maturity index of plant-parasitic nematodes, (B) Shannon trophic diversity index, and (C) fungal to bacterial feeding nematodes ratio for annual (solid line) and perennial (dashed line) crops sampled in North Carolina. Dotted lines represent median values. (From Neher, D.A. and Campbell, C.L., Nematode communities and microbial biomass in soils with annual and perennial crops, *Appl. Soil Ecol.*, 1, 17–28, 1994. With permission.)

second experiment compared effects of 4 different farming systems (manure only [O]; mineral fertilizer only [F]; mineral fertilizer plus herbicides [HF]; and mineral fertilizer plus herbicides plus insecticides [HFI] near Ithaca, Nebraska) (Neher and Olson, 1999).

In the first experiment, values of PPI were greater in soils managed organically than they were in those managed conventionally, but no differences in MI were observed between farming systems (see Figure 7.4). In the second experiment, successional maturity did not differ among management systems, but populations of early successional taxa (weights = 1–2) were greater in the higher input farming systems, suggesting that soils amended with mineral fertilizer and pesticides were at a less mature stage than soils amended with organic matter only. Contrary to predictions, trophic diversity was greatest in the HFI system and least in the O system (Neher and Olson, 1999).

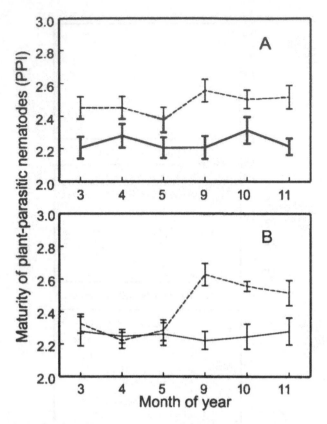

**Figure 7.4**   Mean values of the plant parasitic maturity index (PPI) in soils managed organically (dashed line) and conventionally (solid line) in spring (3 = March, 4 = April, 5 = May) and fall (9 = September, 10 = October, 11 = November) of (A) 1993 and (B) 1994. Standard error bars are illustrated. (From Neher, D.A., Nematode communities in organically and conventionally managed agricultural soils, *J. Nematol.*, 31, 142–154, 1999b.)

Results from both experiments may reflect similar frequencies of disturbance in all management systems. For example, soils managed organically underwent frequent cultivation. The cultivation was replaced by the use of herbicides in soils managed conventionally. Cultivation decreases maturity and trophic diversity index values (Neher and Campbell, 1994). Based on the similarity of index values for the different soil management practices, it is possible to conclude that organically managed systems are not useful as a reference base for maturity indices for annual crops in conventionally managed soils (Neher, 1999b). Physical disturbances such as cultivation disturb soil nematode community structure and function as much as, or more than, applications of synthetic chemicals such as fertilizers and pesticides (Neher and Campbell, 1994; Neher, 1995). Sites with minimal to no physical disturbance may serve as better reference sites in environmental monitoring programs than sites without application of synthetic chemicals (Neher, 1999b).

### 7.5.3  Confounded Disturbance Regimes

As suggested above, a major challenge in assessing soil quality is the confounded nature of physical and chemical disruption in most management practices. Some taxa may be sensitive to one or both types of disturbance, with similar or inverse associations (Fiscus, 1997). Fiscus hypothesizes that physical changes in soil can act as a disturbance while chemical/nutrient changes can act as an enrichment, leading to qualitatively different impacts on nematode communities. Indirect impacts of disturbance on nematode community composition often exceed direct impacts. Greater sensitivity to indirect effects suggests that nematode communities are more responsive to the secondary impacts (those mediated by the soil environment) of agricultural management than the impacts of cultivation or chemical/nutrient applications themselves (Fiscus, 1997).

## 7.6  REGIONAL SCALE APPLICATION

The goal of contemporary environmental monitoring programs is to compare nematode communities in soils between geographic regions ranging in size from 375,000 to 600,000 $km^2$ (Neher et al., 1998). The author participated in a multiple region survey to determine the feasibility of using maturity and trophic diversity indices of nematode communities on large geographic scales. In these surveys, soil samples were collected from fields that were planted with annual row crops or an annually harvested perennial crop such as hay or alfalfa in North Carolina (n = 164) and Nebraska (n = 154) (Neher et al., 1998). Fields were selected using an area frame design (Cotter and Nealon, 1987). Within every field, soils were sampled by taking one core at each of 20 equally spaced sites along one 90-m linear transect across a random 2-ha area to obtain data to measure variation among fields (Neher et al., 1995; 1998). Cores for each transect were mixed thoroughly by hand to form a composite sample to reduce variance associated with the aggregated spatial pattern of nematodes in soil (Barker and Campbell, 1981) and obtain a realistic representation of the nematode community in the field. In every sixth field, a second transect was also sampled to quantify variability within fields. Variability within composite samples was quantified by splitting composite soil samples of double volume taken from a second independent transect in every twelfth field (Neher and Campbell, 1996; Neher et al., 1998).

Modifications of maturity indices increased reliability (signal to noise ratio) and thus improved performance on a regional scale (Neher and Campbell, 1996). In one modification, values for the early colonizing taxa (weight = 1) (Popovici, 1992) were removed from the original MI index (Bongers, 1990) to give a new index, MI25 (Bongers et al., 1995; described as "MINO" in Neher and Campbell, 1996). In the other modification, plant feeding and free living taxa of nematodes were combined in a single index, $\Sigma MI$ (Yeates, 1994). Relatively large total and within sample variance for the original MI index compared to the modified MI25 index suggest that inclusion of the free living opportunists (weight = 1) decreases the reliability of detection by inflating the noise factor (Neher and Campbell, 1996).

Differences in successional maturity and trophic diversity of nematode communities between North Carolina and Nebraska, two states of contrasting climates and soil types, exceeded those among Land Resource Regions (LRRs) within states. LRRs represent geographic areas with unique soil type, topography, climate, and water resources (USDA-SCS, 1981). These relatively large differences suggest that maturity and trophic diversity indices of soil quality need not be calibrated independently when applied in geographic areas smaller than the land area of North Carolina (126,180 km$^2$) and Nebraska (199,120 km$^2$) (Neher and Campbell, 1996). It has yet to be determined how large a region or state can be before subdivision is necessary for independent calibrations of indices.

Based upon the results of the regional surveys, it appears that regional studies require a minimum sampling of 50 to 100 fields, with three independent samples (transects) per field and two subsamples assayed per sample. If cost is a major limiting factor, an alternative would be to sample a larger number of fields with only one subsample for each of two samples assayed per field (Neher and Campbell, 1996). The information obtained using this option would have a smaller degree of reliability; however, cost is often a driving factor in sampling programs. For states or regions such as North Carolina, in which plant feeding nematodes are major agricultural pests, an index such as MI or MI25 may be a better choice than PPI if the focus of the study is to examine overall maturity or stability of nematode communities. For states or regions such as Nebraska, in which plant feeding nematodes occur but are less prevalent, ecological indices that include plant parasitic nematodes, such as PPI and ΣMI, may be the better choices because they indicate variability among fields more reliably than indices that only include free living nematodes (Neher and Campbell, 1996).

## 7.7 FUTURE RESEARCH

Many challenges remain to be overcome before it is possible to fully understand and interpret maturity and trophic diversity indices.

- First, our ability to resolve trophic or functional groups must be improved, because it limits our current understanding. Many species have yet to be assigned to trophic or functional groups (Brussaard et al., 1997). Ultimately, resolution at a species level is desirable (Neher, 1999a). However, before this is possible, experiments must be conducted to learn the natural history traits of free living species and the response of each species to various types of environmental stresses. These experiments will help us to understand and identify which species or groups of species have key functions in the maintenance of energy and material flow through an ecosystem (de Bruyn, 1997).
- Second, a more thorough understanding of the sequence of community succession relative to soil function dynamics would be useful in establishing the kinds of community composition associated with ecologically sound agricultural systems (Neher, 1999a). These associations can only be revealed with appropriate sampling schedules that consider the lag periods that may occur between nematode population peaks and subsequent ecological process change.

- Third, quantitative associations that reveal cause/effect relationships or mechanisms between nematodes and ecosystem function are necessary for complete understanding of indicator performance.
- Fourth, alternative indices for describing how nematode communities respond to environmental stress must be developed and evaluated. For example, an alternative to the maturity index is an index based on reproductive, gender determination, and dispersal traits (Siepel, 1995). The index would be sensitive to how these characteristics vary among the different stages of the nematode life cycle and how the dominant versions of these traits change with different types and frequencies of disturbance and stress.

## REFERENCES

Allison, F.E., Soil organic matter and its role in crop production. *Dev. Soil Sci.*, Vol. 3, Elsevier, Amsterdam, 1973.

Anderson, R.V., Gould, W.D., Woods, L.E., Cambardella, C., Ingham, R.E., and Coleman, D.C., Organic and inorganic nitrogenous losses by microbivorous nematodes in soil, *Oikos*, 40, 75–80, 1983.

Barker, K.R. and Campbell, C.L., Sampling nematode populations, in Zuckerman, B.M., and Rohde, R.A., Eds., *Plant Parasitic Nematodes*, Academic Press, NY, Vol. 3, 51–474, 1981.

Beare, M.H., Fungal and bacterial pathways of organic matter decomposition and nitrogen mineralization in arable soil, in Brussaard, L. and Ferrera-Cerrato, R., Eds., *Soil Ecology in Sustainable Agricultural Systems*, Lewis Publishers, Boca Raton, FL, 37–70. 1997.

Bolton, H., Jr. Soil microbial biomass and selected soil enzyme activities on an alternatively and a conventionally managed farm, M.S. thesis, Washington State University, Pullman, 1983.

Bongers, T., The maturity index: an ecological measure of environmental disturbance based on nematode species composition, *Oecologia*, 83, 14–19, 1990.

Bongers, T., Alkemade, R., and Yeates, G.W., Interpretation of disturbance-induced maturity decrease in marine nematode assemblages by means of the Maturity Index, *Mar. Ecol. Prog. Ser.*, 76, 135–142, 1991.

Bongers, T., de Goede, R.G.M., Korthals, G.W., and Yeates, G.W., Proposed changes of c-p classification for nematodes, *Russian J. Nematol.*, 3, 61–62, 1995.

Brussaard, L., Behan-Pelletier, V.M., Bignell, D.W., Brown, V.K., Didden, W., Folgarait, P., Fragoso, C., Freckman, D.W., Gupta, V.V.S.R., Hattori, T., Hawksworth, D.L., Klopatek, C., Lavelle, P., Malloch, D.W., Rusek, J., Söderström, B., Tiedje, J.M., and Virginia, R.A., Biodiversity and ecosystem functioning in soil, *Ambio*, 26, 563–570, 1997.

Cotter, J., and Nealon, J., Area frame design for agricultural surveys, USDA, National Agricultural Statistics Service, Research and Applications Division, Washington, D.C., 1987.

de Bruyn, L.A.L., The status of soil macrofauna as indicators of soil health to monitor the sustainability of Australian agricultural soils, *Ecol. Econ.*, 23, 167–178, 1997.

de Goede, R., *Terrestrial Nematodes in a Changing Environment*, CIP-gegevens Koninkljke Bibliotheek, Den Haag, Wageningen, The Netherlands, 1993.

Dix, N.J. and Webster, J., *Fungal Ecology*, Chapman and Hall, London, UK, 1995.

Elliott, E.T., Hunt, H.W., and Walter, C.E., Detrital food web interactions in North American grassland ecosystems, *Agric. Ecosystems Environ.*, 24, 41–56, 1988.

Ettema, C.H. and Bongers, T., Characterization of nematode colonisation and succession in disturbed soil using the Maturity Index, *Biol. Fertil. Soils*, 16, 79–85, 1993.

Ferris, V.R. and Ferris, J.C., Inter-relationships between nematode and plant communities in agricultural ecosystems, *Agro-Ecosystems*, 1, 275–299, 1974.

Fiscus, D.A., Development and evaluation of an indicator of soil health based on nematode communities, M.S. thesis, North Carolina State University, Raleigh, NC, 1997.

Freckman, D.W., Bacterivorous nematodes and organic-matter decomposition, *Agric. Ecosystems Environ.*, 24, 195–217, 1988.

Freckman, D.W. and Ettema, C.H., Assessing nematode communities in agroecosystems of varying human intervention, *Agric. Ecosystems Environ.*, 45, 239–261, 1993.

Giller, K.E., Beare, M.H., Lavelle, P., Izac, A.M.N., and Swift, M.J., Agricultural intensification, soil biodiversity and agroecosystem function, *Appl. Soil Ecol.*, 6: 3–16, 1997.

Hassink, J., Bouwman, L.A., Zwart, K.B., and Brussaard, L., Relationships between habitable pore space soil biota and mineralization rates in grassland soils, *Soil Biol. Biochem.*, 25, 47–55, 1993.

Hawksworth, D.L. and Mound, L.A., Biodiversity databases: the crucial significance of collections, in Hawksworth, D.L., Ed., *The Biodiversity of Microorganisms and Invertebrates: Its Role in Sustainable Agriculture*, CAB International, Wallingford, UK, 1991, 17–29.

Hendrix, P.F., Parmelee, R.W., Crossley, D.A., Jr., Coleman, D.C., Odum, E.P., and Groffman, P.M., Detritus food webs in conventional and no-tillage agroecosystems, *BioScience*, 36, 374–380, 1986.

Hunt, H.W., Coleman, D.C., Ingham, E.R., Ingham, R.E., Elliott, E.T., Moore, J.C., Rose, S.L., Reid, C.P.P., and Morley, C.R., The detrital food web in a shortgrass prairie, *Biol. Fertil. Soils*, 3, 57–68, 1987.

Hyvonen, R. and Persson, T., Effects of acidification and liming on feeding groups of nematodes in coniferous forest soils, *Biol. Fertil. Soils*, 9, 205–210, 1990.

Ingham, R.E., Trofymow, J.A., Ingham, E.R., and Coleman, D.C., Interactions of bacteria, fungi, and their nematode grazers: Effects on nutrient cycling and plant growth, *Ecol. Monogr.*, 55, 119–140, 1985.

Lubchenco, J., Olson, A.M., Brubaker, L.B., Carpenter, S.R., Holland, M.M., Hubbell, S.P., Levin, S.A., MacMahon, J.A., Matson, P.A., Melillo, J.M., Mooney, H.A., Peterson, C.H., Pulliam, H.R., Real, L.A., Regal, P.J., and Risser, P.G., The sustainable biosphere initiative: an ecological research agenda, *Ecology*, 72, 371–412, 1991.

Ludwig, J.A. and Reynolds, J.F., *Statistical Ecology: A Primer on Methods and Computing*, Wiley, New York, 1988.

Moore, J.C. and P.C. de Ruiter, Temporal and spatial heterogeneity of trophic interactions within below-ground food webs, *Agric. Ecosystems Environ.*, 34: 371–397, 1991.

Moore, J.C., Walter, D.E., and Hunt, H.W., Arthropod regulation of micro- and mesobiota in below-ground detrital food webs, *Annu. Rev. Entomol.*, 33, 419–439, 1988.

Neher, D.A., Biological diversity in soils of agricultural and natural ecosystems, in Olson, R.K., Francis, C.A., and Kaffka, S., Eds., *Exploring the Role of Diversity in Sustainable Agriculture*, American Society of Agronomy, Madison, WI, 1995, 55–72.

Neher, D.A., Soil community composition and ecosystem processes: Comparing agricultural ecosystems with natural ecosystems, *Agroforestry Syst.*, 45, 159–185, 1999a.

Neher, D.A., Nematode communities in organically and conventionally managed agricultural soils, *J. Nematol.*, 31, 142–154, 1999b.

Neher, D.A. and Campbell, C.L., Nematode communities and microbial biomass in soils with annual and perennial crops, *Appl. Soil Ecol.*, 1, 17–28, 1994.

Neher, D.A. and Campbell, C.L., Sampling for regional monitoring of nematode communities in agricultural soils, *J. Nematol.*, 28, 196–208, 1996.

Neher, D.A. and Olson, R.K., Nematode communities in soils of four farm cropping management systems, *Pedobiologia*, 43, 430–438, 1999.

Neher, D.A., Peck, S.L., Rawlings, J.O., and Campbell, C.L., Measures of nematode community structure for an agroecosystem monitoring program and sources of variability among and within agricultural fields, *Plant Soil*, 170, 167–181, 1995.

Neher, D.A., Easterling, K.N., Fiscus, D., and Campbell, C.L., Comparison of nematode communities in agricultural soils of North Carolina and Nebraska, *Ecol. Appl.*, 8, 213–223, 1998.

Popovici, I., Nematodes as indicators of ecosystem disturbance due to pollution, *Stud. Univ. Babes-Bolyai Ser. Biol.*, 37, 15–27, 1992.

Samoiloff, M.R., Nematodes as indicators of toxic environmental contaminants, in Veech, J.A. and Dickson, D.W., Eds., *Vistas on Nematology*, Society of Nematologists, Hyattsville, MD, 1987, 433–439.

Seastedt, T.R., The role of microarthropods in decomposition and mineralization processes, *Annu. Rev. Entomol.*, 29, 25–46, 1984.

Seastedt, T.R., James, S.W., and Todd, T.C., Interactions among soil invertebrates, microbes and plant growth in the tallgrass prairie., *Agric. Ecosystems Environ.*, 24, 219–228, 1988.

Shannon, C.E. and Weaver, W., *The Mathematical Theory of Communication*, University of Illinois, Urbana, 1949.

Siepel, H., Applications of microarthropod life-history tactics in nature management and ecotoxicology, *Biol. Fertil. Soils*, 19, 75–83, 1995.

Simpson, E.H., Measurement of diversity, *Nature*, 163, 688, 1949.

Sohlenius, B. and Sandor, A., Vertical distribution of nematodes in arable soil under grass (*Festuca pratensis*) and barley (*Hordeum distichum*), *Biol. Fertil. Soils*, 3, 19–25, 1987.

Sohlenius, B., Persson, H., and Magnusson, C., Distribution of root and soil nematodes in a young Scots pine stand in central Sweden, *Ecol. Bull.* (Stockholm), 25, 340–347, 1977.

Sohlenius, B., Böstrom, S., and Sandor, A., Carbon and nitrogen budgets of nematodes in arable soil, *Biol. Fertil. Soils*, 6, 1–8, 1988.

Trofymow, J.A. and Coleman, D.C., The role of bacterivorous and fungivorous nematodes in cellulose and chitin decomposition in the context of a root/rhizosphere/soil conceptual model, in Freckman, D.W., Ed., *Nematodes in Soil Ecosystems*, University of Texas, Austin, 1982, 117–138.

United States Department of Agriculture, Report and recommendations on organic farming, U.S. Government Printing Office, Washington, D.C., 1980.

USDA-SCS, Land resource regions and major resource areas of the United States, USDA-SCS Agriculture Handbook 296, U.S. Government Printing Office: Washington, D.C., 1981.

Verhoef, H.A. and Brussard, L., Decomposition and nitrogen mineralization in natural and agro-ecosystems. The contribution of soil animals, *Biogeochemistry*, 11, 175–211, 1990.

Vinciguerra, M.T., Role of nematodes in the biological processes of the soil, *Bull. Zool.*, 46, 363–374, 1979.

Wasilewska, L., The structure and function of soil nematode communities in natural ecosystems and agrocenoses, *Pol. Ecol. Stud.*, 5, 97–145, 1979.

Wasilewska, L., Impact of human activities on nematodes, in Charholm, C., and Bergström, L., Eds., *Ecology of Arable Land*, Kluwer Academic, Dordrecht, The Netherlands, 1989, 123–132.

Wasilewska, L., The effect of age of meadows on succession and diversity in soil nematode communities, *Pedobiologia*, 38, 1–11, 1994.

Wasilewska, L., Jakubczyk, H., and Paplinska, E., Production of *Aphelenchus avenae* Bastian (Nematoda) and reduction of mycelium of saprophytic fungi by them, *Pol. Ecol. Stud.*, 1, 61–73, 1975.

Whitford, W.G., Freckman, D.W., Santos, P.F., Elkins, N.Z., and Parker, L.W., The role of nematodes in decomposition in desert ecosystems, in Freckman, D.W., Ed., *Nematodes in Soil Ecosystems*, University of Texas, Austin, 1982, 98–115.

Yeates, G.W., How plants effect nematodes, *Adv. Ecol. Res.*, 17, 61–113, 1987a.

Yeates, G.W., Significance of developmental stages in the co-existence of three species of Mononchoidea (Nematoda) in a pasture soil, *Biol. Fertil. Soils*, 5, 225–229, 1987b.

Yeates, G.W., Modification and qualification of the nematode maturity index, *Pedobiologia*, 38, 97–101, 1994.

Yeates, G.W. and Coleman, D.C., Nematodes in decomposition, in Freckman, D.W., Ed., *Nematodes in Soil Ecosystems*, University of Texas, Austin, 1982, 55–80.

Yeates, G.W. and Wardle, D.A., Nematodes as predators and prey: relationships to biological control and soil processes, *Pedobiologia*, 40, 43–50, 1996.

Yeates, G.W., Bongers, T., de Goede, R.G.M., Freckman, D.W., and Georgieva, S.S., Feeding habits in soil nematode families and genera: An outline for soil ecologists, *J. Nematol.*, 25, 315–331, 1993.

# CHAPTER 8

# Field-Scale Nutrient Cycling and Sustainability: Comparing Natural and Agricultural Ecosystems

Joji Muramoto, Erle C. Ellis, Zhengfang Li, Rodrigo M. Machado, and Stephen R. Gliessman

## CONTENTS

## 8.1 INTRODUCTION

Nutrient cycling — the movement of nutrients from the biotic components of ecosystems to their abiotic components and back again — is an essential function of all ecosystems (Odum, 1969). Nutrient cycles are never completely closed; all systems receive some external inputs (e.g., from precipitation) and experience some losses to the outside environment (e.g., through denitrification).

Compared to natural ecosystems in general, agroecosystems have relatively open nutrient cycles because they are designed to produce nutrient rich products that are harvested and removed from the system. Many agroecosystems are more leaky than need be. The tillage and exposure of soils between cropping seasons can cause large nutrient losses through leaching and erosion, contaminating ground and surface waters, and requiring large fertilizer inputs to maintain yields (Cox and Atkins, 1979; Odum, 1969).

The more closed an agroecosystem's nutrient cycle, the more it will resemble natural ecosystems and will be ecologically sustainable over the long-term (Ewel, 1999; Jarrell, 1990; Woodmansee, 1984). In designing sustainable agroecosystems, it is useful to compare indicators of nutrient cycling between different management patterns of agroecosystems and natural ecosystems to find systems with more closed nutrient cycles (Gliessman, 1990; 1998).

This chapter explores the use of indicators, focusing on the Cycling Index (CI), a relative measure of material cycling derived from flow analysis (Finn, 1976). Our purpose is to assess the use of the CI and its associated cycling measures as field scale indicators of ecological sustainability in agroecosystems. We do this using three case studies of nitrogen cycling: organic and conventional monocropped strawberry systems in California (Gliessman, et al., 1996), organic and conventional strawberry-paddy rice double-cropping systems in Nanjing, China (Li, et al., 1998), and a natural chaparral ecosystem in California (Grey, 1982).

## 8.2 FLOW ANALYSIS AND THE SOIL-PLANT MODEL

Flow analysis, also known as input–output analysis, is a set of methods used to analyze flows within compartmental models (Hannon, 1973; Ulanowicz, 1986). Based on flow analysis, Finn (1976, 1978, 1980, 1982) developed a set of indicators to describe different aspects of nutrient cycling in ecosystems, including the CI.

The CI is a whole system property, defined as "the fraction of total system throughflow that is cycled," and has been used to describe nutrient cycling in natural ecosystems (Finn, 1978; Patten and Finn, 1979; Christensen, 1995; Han, 1997) and in agricultural ecosystems on the farm scale (Luo and Lin, 1991; Fores and Christian, 1993; Dalsgaard and Oficial, 1997). Because CI is sensitive to model structure (Finn, 1976), we use a standardized two-pool soil–plant model for comparing soil-plant nutrient cycling across agroecosystems and natural ecosystems (Figure 8.1). The model consists of two nutrient pools (soil and plant) and six flows. Pool size represents the change over time in the magnitude of the pool ($\Delta$ plant pool, $\Delta$ soil pool). Each flow is expressed as the sum of component flows. Figure 8.2, which illustrates nitrogen flows and their components in the case studies, is an example of the use of this model.

Although the two-pool system is a vastly oversimplified model of ecosystems, it can be prepared for virtually any terrestrial ecosystem and serves as a convenient reference standard for quantitative comparisons between ecosystems.

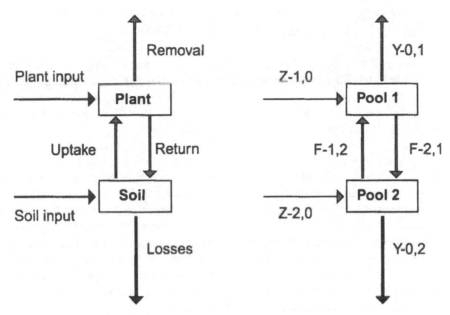

**Figure 8.1** Standardized two-pool soil-plant model, showing conceptual model (left) and formal description (right).

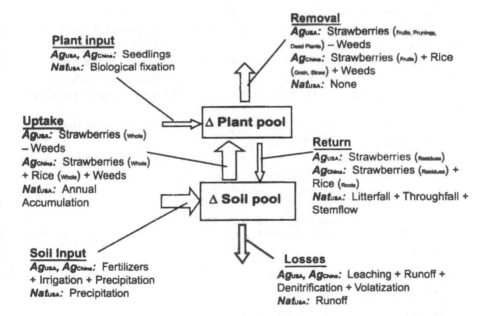

**Figure 8.2** Components of nitrogen flows in natural and agricultural ecosystems in California and in Nanjing. $Ag_{USA}$: Strawberry agroecosystem in California. $Ag_{China}$: Strawberry-paddy rice agroecosystem in Nanjing. $Nat_{USA}$: Chaparral natural ecosystem in California.

## 8.3 ORGANIC AND CONVENTIONAL STRAWBERRY SYSTEMS IN CALIFORNIA AND IN NANJING

We use on-farm field experiments comparing organic and conventional strawberry systems in California (Gliessman et al., 1996) and in Nanjing, China (Li et al., 1998), as case studies in agroecosystem nutrient cycling. These experiments were conducted over 3-year periods to investigate yield and ecosystem process responses to alternative farming practices.

Under a Mediterranean climate, strawberries are grown as annual crops off the central coast of California. They have a 5- to 6-month harvest period and are the most economically valuable crops in the region. The temperate, sunny climate, the long rain free bearing season, and the sandy loam soils of California's coastal valleys promote possibly the heaviest harvests of strawberries in the world (Processing Strawberry Advisory Board, 1989, cited by Wells, 1996). Strawberries represent a high value specialty crop with exacting cosmetic standards. They are one of the most input-intensive field crops in California, involving soil fumigation, plastic mulch, irrigation, and concentrated semi permanent hand labor in all production phases. Acreage planted for certified organic strawberries on the central coast is on the rise, due to increased demand and higher profit margins.

The California experiment was conducted in Davenport from September 1987 to August 1990. The organic plot was managed according to the guidelines of the California Certified Organic Farmers, where conventional plot management conformed to local farm adviser's recommendations.

The strawberry–paddy rice system in Nanjing represents an intensive wet rice system common in many countries in Asia. Crop yields, especially those of rice, were quite high, as were the rates of fertilizer application. The Nanjing experiment was conducted from October 1992 to October 1995. We applied animal manure (pig and cow) and biogas sludge for strawberries in the organic plot and chemical fertilizers in the conventional plot; chemical fertilizers were applied in both plots for rice.

Table 8.1 describes the two study sites, and Table 8.2 summarizes the fertilizer application regimes used in each.

Strawberry yield was much higher in California than in Nanjing. Organic system yields in California were depressed relative to conventional system yields by 39% in the first year, 30% in the second, and 28% in the third (Gliessman et al., 1996). Organic

**Table 8.1   Characteristics of the Study Sites**

|  | USA | China |
|---|---|---|
| Site | Davenport, California | Zhujiang, Nanjing |
| Climate Zone | Mediterranean | Subtropical monsoon |
| Yearly Mean Temperature | 13°C | 16°C |
| Mean Annual Precipitation | 760 mm | 1106 mm |
| Soil Type | Pinto loam (typic argixeroll) | Brown loam |
| Cropping System | Strawberry mono cropping | Strawberry–rice double cropping |

Table 8.2 Fertilizer Application for the Study Sites (Third Year)

| Site | Crop | Conventional plot Fertilizer | kg N/ha | Organic plot Fertilizer | kg N/ha |
|---|---|---|---|---|---|
| USA | Strawberries | Controlled release fertilizer | 102 | Commercial compost Bloodmeal, bone meal | 220 |
| | | total N | 102 | total N | 220 |
| China | Strawberries | Chemical fertilizers[a] | 408 | Pig and cow manure Plant ash, biogas sludge | 325 |
| | Rice | Chemical fertilizers[a] | 215 | Chemical fertilizers[a] | 215 |
| | | total N | 623 | total N | 540 |

[a] Compound fertilizer, potassium chloride, ammonium bicarbonate, and urea were used.

plots had higher yields of rice and strawberries compared with the conventional plots and maintained better soil quality over three years in Nanjing (Li et al., 1998).

To demonstrate our methods for nutrient cycling analysis, we use nitrogen balance data from the third year of each experiment. Crop yields for the third year of each experiment are shown in Table 8.3. Nutrient cycling data for chaparral scrub ecosystems (Grey, 1982) are used as a benchmark for comparison with the California agroecosystem experiments; chaparral scrub is one of the representative natural ecosystems on the central coast of California.

Figure 8.2 summarizes the nitrogen flows and pools that we identified in the three case studies.

## 8.4 OBSERVATIONAL UNCERTAINTY ANALYSIS

To conduct a statistically reliable analysis of nitrogen flow, we use an observational uncertainty analysis system based on methods described by Ellis et al., 2000. This system uses probability distribution functions (PDFs) to describe our degree of belief, or betting odds, for the mean of every variable measured or estimated in this study. Monte Carlo methods are used to estimate probability densities for variables, such as Cycling Indexes, that are calculated as functions of other variables. For most variables, lognormal PDFs were used to avoid negative values unless otherwise stated.

Table 8.3 Third Year Yield of Strawberries and Rice

| Site | Crop | Yield (tons/ha) Conventional plot | Organic plot |
|---|---|---|---|
| USA | Strawberries[a] | 56.3[b] | 40.3[b] |
| China | Strawberries[a] | 10.7 | 12.0 |
| | Rice (grain) | 8.3[c] | 9.0[c] |

[a] Fresh yield.
[b] Marketable fruit yield.
[c] Average of three years.

For U.S. strawberry fields, PDFs for uptake, removal, return, and plant input were calculated from direct measurements or regressions of directly measured data (Hunt, 1982). Fertilizer PDFs were derived by combining direct measurements, guaranteed analysis values, supplier's analytical records, and data from references (Jones, 1979; Soil Improvement Committee, 1995). PDFs of losses from soil (denitrification, volatilization, leaching, and runoff) were estimated based on Meisinger and Randall (1991), Huntley et al. (1997), and Smith and Cassel (1991), using beta-subjective PDFs (Palisade Corporation, 1996). PDFs for precipitation deposition were estimated from monthly precipitation in the central coast region (California Department of Water Resources, 1998) and inorganic nitrogen ($NH_4$-N + $NO_3$-N) concentration in precipitation in California (The National Atmospheric Deposition Program/National Trends Network, 1998) during the experiment period. PDFs for nitrogen loading from irrigation were estimated from the analysis record of the water sources (Santa Cruz Water Quality Laboratory, personal communication) and the farm irrigation record. We standardized the data with a growth period weighted annual average.

For the Chinese strawberry–paddy rice system, PDFs of uptake, removal, return, plant input (seedlings), and soil input (fertilizers) were calculated from direct measurements. PDFs of deposition through precipitation were estimated based on Lu and Shi (1979); nitrogen loading from irrigation and losses from soils (leaching, runoff, denitrification, and volatilization) were estimated based on Xi (1986). Since no variability data was available for the Chinese field experimental data, we used PDFs with sigmas set to a CV of 30% for direct measured variables and 50% for the other variables.

Nitrogen flow data for the undisturbed chaparral natural ecosystem were collected from references. We used data from Grey (1982), Mooney and Rundell (1979), and Marion et al. (1980) for plant uptake and return, from Schlesinger and Grey (1982) for plant input (biological fixation), and for soil output (runoff). To be conservative, we used a 50% CV, or a range (minimum × 0.5, maximum × 1.5) to calculate the PDFs of nitrogen flows in chaparral.

$\Delta$ plant and $\Delta$ soil pools were calculated by subtracting the outflows from the inflows of each pool. Results are presented as means and 90% credible intervals from 10,000 Monte Carlo simulation iterations (CIN; Morgan and Henrion, 1990; Ellis et al., 2000).

## 8.5 NITROGEN FLOWS AND POOLS

Figure 8.3 shows the mean and CIN (5%, 95%) of nitrogen flows and $\Delta$ plant/soil pools of the ecosystems analyzed (kg N/ha/yr). This figure shows distinctive differences in nitrogen flows across the ecosystems.

In California*, the conventional system (in which controlled release fertilizers were applied and the soil fumigated) was highly efficient in nitrogen use:

---

* In the third year of the California strawberry experiment, we used the strawberry plants planted in the second year. Thus, no seedlings were brought into the systems. At the end of the third year, all plants were incorporated into the soil in both the organic and conventional systems. Hence, $\Delta$ plant pool was negative regardless of management practice in the California experiment.

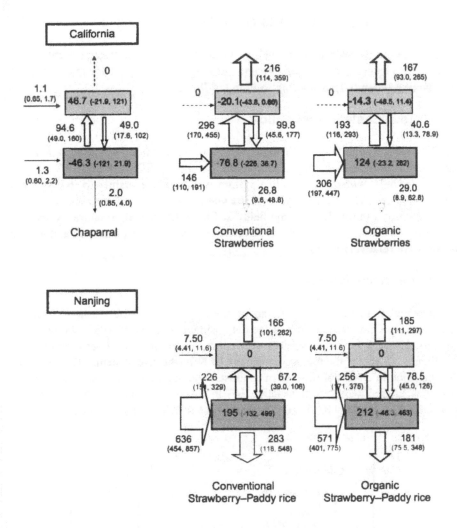

**Figure 8.3** Mean and CIN (5%, 95%) of nitrogen flows and Δ pools (kg N/ha/yr) in natural and agricultural ecosystems in California and in Nanjing.

inputs were lower than those of the organic plot (about half), the harvest was higher (28% higher than the organic plot), more N was returned to the soil, and N losses were slightly lower. However the mean Δ soil pool in the conventional system was negative, suggesting that soil nitrogen was mined from the soil by the plants.

Compared to the California strawberry systems, the strawberry–paddy rice system in Nanjing had greater soil input (~600 kg N/ha/yr) and higher N losses (180 to 280 kg N/ha/yr as means). Most of the N loss is believed to have occurred during the paddy rice period in gaseous form.

The natural chaparral ecosystem had very small inflows and outflows to the environment. Importantly, there was no removal of N in the form of harvest.

## 8.6 INDICATORS OF NUTRIENT CYCLING

To describe and compare nutrient cycling status across the ecosystems, we calculated three system level indicators: total system throughflow (TST), net accumulation, and the Cycling Index (CI), according to Finn (1978, 1982).

### 8.6.1   Total System Throughflow

Total system throughflow (TST) is the sum of all throughflows in the system and is an indicator of system activity (Finn, 1980). As shown in Figure 8.4, TST was the highest in the strawberry-paddy rice system in China due mainly to its large soil input (~600 kg N/ha/yr). Both types of strawberry fields in California had TSTs that were about two thirds those of the fields in China. Chaparral, a natural ecosystem in California, had the lowest TST. Differences in TST between practices at the same locale were relatively small.

### 8.6.2   Net Accumulation

Net accumulation, derived by subtracting the sum of the outputs from the sum of the inputs, is a modified version of the output/input ratio (Finn, 1982; Vitousek and Reiners, 1975). It indicates whether the system is a net loser or a net accumulator.

Figure 8.5 shows the net accumulations of N in the five systems. The mean net accumulation in chaparral was ~0. In contrast, mean net accumulations in the agroecosystems differed widely, from negative (–97 in the conventional strawberry system in California) to positive (110 to 220 in the other systems). The negative net accumulation

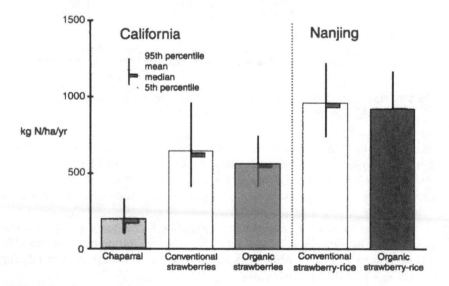

**Figure 8.4**   Total system throughflow (TST) of nitrogen in natural and agricultural ecosystems in California and in Nanjing.

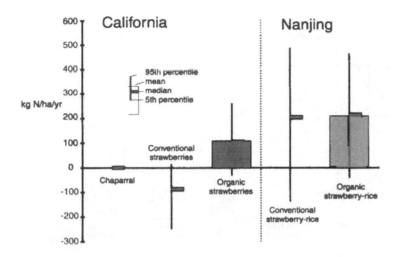

**Figure 8.5**   Net accumulations of nitrogen in natural and agricultural ecosystems in California, and in Nanjing.

in the conventional strawberry system in California indicates that nitrogen was mined from the soil. This might have resulted from the enhancement of soil nitrogen mineralization (Rovira, 1976) and of root development due to root disease control (Yuen, et al., 1991) brought about by the soil fumigation. These net accumulation values are very uncertain. For example, Monte Carlo simulations yield a 7 to 13% probability that the net accumulations of the California organic and Chinese agroecosystems are actually negative. Moreover, there is a ~13% probability that net accumulation in the California conventional strawberry system is positive and a ~8% probability that this accumulation is actually greater than that of the Chinese conventional system. Clearly, the calculation of soil and ecosystem nutrient balances by subtracting outputs from inputs is an uncertain endeavor that can yield misleading results.

## 8.6.3   Cycling Index

The Cycling Index (CI) is a measure of the amount of material cycled relative to the total amount moving within and through the system. It varies from 0 (no cycling) to 1 (all material is cycled).

As shown in Figure 8.6, the highest mean CI (0.49) was recorded for California chaparral. Among the agroecosystems, the highest CI (0.29) belongs to the California conventional system, in which greater quantities of plant residues were returned to the soil compared to the organic system (CI = 0.11). Mean CIs in the Chinese organic and the Chinese conventional strawberry-paddy rice systems were 0.12 and 0.09, respectively. The higher the mean CI, the greater the uncertainty of the CI.

To examine the relationship between the amount of plant residue returned to the systems and CI, we conducted three scenario analyses (Table 8.4). In scenario A,

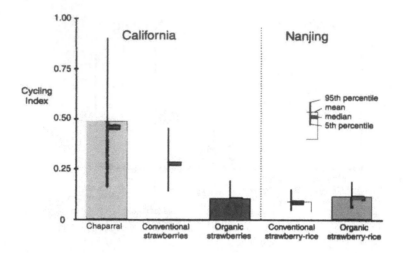

**Figure 8.6**  The Cycling Index (CI) for nitrogen in natural and agricultural ecosystems in California and Nanjing.

no residue is returned. In scenario B, crop residues are returned (California: strawberry residues, except prunings; Nanjing: strawberry residues plus rice roots). In scenario C, both crop residues and weeds are returned (California: scenario B plus weeds; Nanjing: scenario B plus rice straw plus weeds). Regardless of practice and site, CI is zero when no plant residues are returned to the system. When plant residues are returned, CI increases with the amount of biomass returned, although the maximum (0.31, for California conventional under scenario C) is still short of the CI for the chaparral ecosystem (0.49).

## 8.7 NUTRIENT CYCLING IN AGROECOSYSTEMS VS. NATURAL ECOSYSTEMS

Our flow analyses indicate that regardless of site and management practice, agroecosystems tend to have high TSTs, a wide range of net accumulation, low CIs, and

**Table 8.4  Cycling Index Values for Three Scenarios of Plant Residue Return**

|  |  | Scenario A (no return) | Scenario B (crop residue) | Scenario C (crop residue + weeds) |
|---|---|---|---|---|
| USA[a] | Conventional | 0.00 | 0.29 | 0.31 |
|  | Organic | 0.00 | 0.11 | 0.14 |
| China[b] | Conventional | 0.00 | 0.09 | 0.14 |
|  | Organic | 0.00 | 0.12 | 0.16 |

[a] For California, Scenario A: No residue return. Scenario B: Strawberry residue return (except pruning). Scenario C: Scenario B plus weeds return.
[b] For Nanjing, Scenario A: No residue return. Scenario B: Strawberry residue plus rice roots return. Scenario C: Scenario B plus rice straw plus weeds return.

Table 8.5  Summary of Nutrient Cycling Status for Natural and
Agricultural Ecosystems

|  | Agroecosystem | Natural Ecosystem |
| --- | --- | --- |
| TST | High | Low |
| Net accumulation | Negative to High | ~ 0 |
| Cycling Index | Low | High |
| Harvest removal | High | None |

high removal through harvest. These values contrast with those of the natural system we analyzed, which has a low TST, approximately zero net accumulation, a high CI, and no harvest. These generalizations, summarized in Table 8.5, agree well with the observation that net accumulation in an long undisturbed ecosystem tends to be close to zero (Vitousek and Reiners, 1975; Finn, 1982).

Although further investigation is required, some important points regarding the Cycling Index can be drawn from this analysis. Although a higher CI may be correlated with greater sustainability, there are clear limits to how far CI can be increased without affecting harvests. A CI approaching that of a natural ecosystem is not desirable, since it would be achieved at the cost of reducing harvests to zero. Practically speaking, there are many constraints to increasing the CI of agroecosystems. Although CI tends to increase when the amount of plant residue returned to the soil is increased (Table 8.4), achieving an increased return contrasts sharply with the general direction of crop breeding, which aims to increase the harvest indices (edible biomass/above ground biomass) of crops (e.g., Loomis and Connor, 1992).

## 8.8 FUTURE STUDIES

Some important questions about nutrient cycling need to be answered by future research: How far we can increase CI without decreasing harvest? What are effective techniques for increasing CI? We need quantitative studies on the relationship between CI and other indices such as crop harvest, nitrogen use efficiency, and nutrient balance. Jarrell (1990) states that, "Efforts should be made to close the N cycle as nearly as possible and to introduce N into the farm in the most efficient manner for long-term production." To accomplish this goal, we need to apply field-scale CI and N-use efficiency as indicators.

Future research should:

- Further test the two-pool model by (1) analyzing the cycling of other elements, (2) applying the model to different farming systems (e.g., inter-cropping systems, cover cropping systems, and agroforestry systems), and (3) applying it to different management practices.
- Base nutrient cycling analyses on time scales rather than annual increments (e.g., monthly).
- Develop an indicator for cycling rate because CI says nothing about the rate of nutrient cycling (Finn, 1978).
- Measure the Δ soil pool directly by comparing soil nutrient content two or more times, and examine the effect of this method on the indicators.

## 8.9 CONCLUSIONS

Flow analysis based on a standardized two pool soil-plant model is a simple and useful method for quantitative comparisons of nutrient cycling across natural ecosystems and agroecosystems. Indicators, including total system throughflow (for system activity), net accumulation (for nutrient balance), and the Cycling Index (for nutrient cycling), can be used to compare different aspects of nutrient cycling in ecosystems. When used with observational uncertainty analysis, these indicators can be compared in a statistically reliable way. The two-pool model is too simplistic to evaluate agroecosystem nutrient cycling at scales larger than that of a field.

The Cycling Index may be a valuable indicator of the ecological sustainability of agroecosystems at the field scale. Comparisons between agroecosystems and a natural ecosystem suggest that a contradiction exists between the degree of nutrient cycling (as measured by the Cycling Index) and crop yield.

## REFERENCES

California Department of Water Resources, Division of Flood Management, URL: http://cdec.water.ca.gov/. Accessed, 5/11/1998.

Christensen, V., Ecosystem maturity: Towards quantification, *Ecological Modelling*, 77, 3–32, 1995.

Cox, G.W., and Atkins, M.D., *Agricultural Ecology*, Freeman, San Francisco, 1979.

Dalsgaard, J.P.T. and Oficial, R.T., A quantitative approach for assessing the productive performance and ecological conditions of smallholder farms, *Agric. Syst.*, 55, 503–533, 1997.

Ellis, E.C., Li, R.G., Yang, L.Z., and Cheng, X., Long-term changes in village-scale ecosystems in China using landscape and statistical methods, *Ecological Appl.*, in press.

Ewel, J.J., Natural systems as models for the design of sustainable systems of land use, *Agroforestry Syst.*, 45, 1–21, 1999.

Finn, J.T., Measures of ecosystem structure and function derived from analysis of flows, *J. Theoretical Biol.*, 56, 363–380, 1976.

Finn, J.T., Ecosystem succession, nutrient cycling and output–input ratios, *J. Theoretical Biol.*, 99, 479–489, 1982.

Finn, J.T., Cycling Index: a general definition for cycling in compartment models, in Adriano, D.C., and Brisbin, I.L.J., Eds., *Environmental Chemistry and Cycling Processes*, Technical Information Center, U.S. Department of Energy, Augusta, Georgia, 1978, 138–206.

Finn, J.T., Flow analysis of models of the Hubbard brook ecosystem, *Ecology*, 61, 562–571, 1980.

Fores, E., and Christian, R.R., Network analysis of nitrogen cycling in temperate wetland ricefields, *Oikos*, 67, 299–308, 1993.

Gliessman, S.R., Ed., *Agroecology: Researching the Ecological Basis for Sustainable Agriculture*, Springer-Verlag, New York, 1990.

Gliessman, S.R., *Agroecology: Ecological Processes in Sustainable Agriculture*, Ann Arbor Press, Chelsea, MI, 1998.

Gliessman, S.R., Werner, M.R., Swezey, S.L., Casswell, E., Cochran, J., and Rosado-May, F., Conversion to organic strawberry management changes ecological processes, *Calif. Agric.*, 50, 24–31, 1996.

Grey, J.T., Comparative nutrient relations in adjacent stands of chaparral and coastal sage scrub, in Conrad, C.E. and Oechel, W.C., Eds., *The Symposium on Dynamics and Management of Mediterranean-type Ecosystems*, USDA Forest Service, San Diego State University, San Diego, 1982, 306–312.

Han, B.P., On several measures concerning flow variables in ecosystems, *Ecological Modelling*, 104, 289–302, 1997.

Hannon, B., The structure of ecosystems, *J. Theoretical Biol.*, 41, 535–546, 1973.

Hunt, R., *Plant Growth Curves; the Functional Approach to Plant Growth Analysis*, Edward Arnold Limited, London, 1982.

Huntley, E.E., Barker, A.V., and Stratton, M.L., Composition and uses of organic fertilizers, in MacKinnon, H.C. and Rechcigl, J.E., Eds., *Agricultural Uses of By-Products and Wastes*, American Chemical Society, Washington, D.C., 1997, 120–139.

Jarrell, W.M., Nitrogen in agroecosystems, in Carroll, C.R., Vandermeer, J.H., and Rosset, P., Eds., *Agroecology*, McGraw-Hill, New York, 1990, 385–411.

Jones, U.S., *Fertilizers and Soil Fertility*, Reston Publishing Company, Reston, VA, 1979.

Li, Z., Xi, Y., Tai, C., and Wang, Q., Balance of soil nutrients in the strawberry-rice system in China, in Patanothai, A., Ed., *Land Degradation and Agricultural Sustainability: Case Studies from Southeast and East Asia*, Khon Kaen University, Khon Kaen, Thailand, 1998, 201–214.

Loomis, R.S., and Connor, D.J., *Crop Ecology: Productivity and Management in Agricultural Systems*, Cambridge University Press, Cambridge, UK, 1992.

Lu, R.K., and Shi, T.J., Nutrient contents in rainfalls in Jinhua area, *Acta Pedologica Sinica*, 16, 81–84, 1979.

Luo, S.M., and Lin, R.J., High bed-low ditch system in the Pearl river delta, south China, *Agric. Ecosystems Environ.*, 36, 101–109, 1991.

Marion, G.M., Kummerow, J., and Miller, P.C., Predicting nitrogen mineralization in chaparral soils, *Soil Sci. Soc. Am. J.*, 45, 956–961, 1980.

Meisinger, J.J. and Randall, G.W., Estimating nitrogen budgets for soil-crop systems, in Follett, R.F., Keeney, D.R., and Cruse, R.M., Eds., Managing Nitrogen for Groundwater Quality and Farm Profitability: Proceedings of a Symposium of the American Society of Agronomy and Soil Science Society of America and Crop Science Society of America, Madison, WI, 1988, 85–124.

Mooney, H.A. and P.W. Rundel, Nutrient relations of the evergreen shrub, *Adenostoma fasciculatum*, in the California chaparral. *Botanical Gazette* 140: 109–113, 1979.

Morgan, M.G. and Henrion, M., *Uncertainty: A Guide to Dealing with Uncertainty in Quantitative Risk and Policy Analysis*, Cambridge University Press, New York, 1990.

Odum, E.P., The strategy of ecosystem development, *Science*, 164, 262–270, 1969.

Palisade Corporation, @RISK advanced risk analysis software for spreadsheets, Palisades Corporation, Newfield, New York. 1996.

Patten, B.C. and Finn, J.T., System approach to continental shelf ecosystems, in Halfon, E., Ed., *Theoretical System Ecology: Advances and Case Studies*. Academic Press, New York, 1979, 183–212.

Rovira, A.D., Studies on soil fumigation, I: Effects on ammonium, nitrate and phosphate in soil and on the growth, nutrition and yield of wheat, *Soil Biol. Biochem.*, 8, 241–247, 1976.

Schlesinger, W.H. and Grey, T., Atmospheric precipitation as a source of nutrients in chaparral ecosystems, in Conrad, C.E. and Oechel, W.C., Eds., *The Symposium on Dynamics and Management of Mediterranean-type Ecosystems*, USDA Forest Service, San Diego State University, San Diego, 1982, 279–284.

Smith, S.J. and Cassel, D.K., Estimating nitrate leaching in soil materials, in Follett, R.F., Keeney, D.R., and Cruse, R.M., Eds., *Managing Nitrogen for Groundwater Quality and Farm Profitability*, Proceedings of a Symposium of the American Society of Agronomy and Soil Science Society of America and Crop Science Society of America, Madison, WI, 1988, 165–188.

Soil Improvement Committee, California Fertilizer Association, *Western Fertilizer Handbook*, Interstate Publishers, Inc., Danville, Illinois, 1995.

The National Atmospheric Deposition Program/National Trends Network, NADP/NTN Monthly Data in California, URL: http://nadp.sws.uiuc.edu/nadpdata/state.asp?state=CA. Accessed 4/30/1998.

Ulanowicz, R.E., *Growth and Development: Ecosystems Phenomenology*, Springer-Verlag, New York, 1986.

Vitousek, P.M. and Reiners, W.A., Ecosystem succession and nutrient retention: a hypothesis, *BioScience*, 25, 376–381, 1975.

Wells, M.J., *Strawberry Fields: Politics, Class, and Work in California Agriculture*, Cornell University Press, Ithaca, New York, 1996.

Woodmansee, R.G. Comparative nutrient cycles of natural and agricultural ecosystems: a step toward principles, in Lowrance, R., Stinner, B.R., and House, G.L., Eds., *Agricultural Ecosystems: Unifying Concepts*, John Wiley & Sons, New York, 1984, 145–156.

Xi, Z.B., N cycling in a farmland ecosystem, *Proceedings of the Soil N Workshop*, The Soil Science Society of China, 1986, 217–227.

Yuen, G.Y., Schroth, M.N., Weinhold, A.R., and Hancock, J.G., Effects of soil fumigation with methyl bromide and chloropicrin on root health and yield of strawberry, *Plant Disease*, 75, 416–420, 1991.

# Section III
# Combining Social and Ecological
# Indicators of Sustainability

CHAPTER 9

# Assessing Agricultural Sustainability Using the Six-Pillar Model: Iran as a Case Study

Abbas Farshad and Joseph A. Zinck

## CONTENTS

## 9.1 INTRODUCTION

A sustainable agricultural system is a system that is politically and socially acceptable, economically viable, agrotechnically adaptable, institutionally manageable, and environmentally sound. Satisfying all these sustainability requirements and the relevant analytical criteria is a complex endeavor; so complex that it may never be implemented for any one system or region. Less comprehensive methods of sustainability assessment, which focus on a particular facet, are more practical to implement but result in greater uncertainty about the overall sustainability of the agroecosystem (Farshad and Zinck, 1993; Zinck and Farshad, 1995).

1-8493-0894-1/01/$0.00+$.50
© 2001 by CRC Press LLC

It is possible to combine comprehensiveness and practicality by conducting more than one type of specific sustainability assessment, and to put these assessments into a conceptual framework that describes what is required for a system to be truly sustainable. This is the approach the authors used in comparing the sustainability of modern and traditional agricultural systems in the Hamadan-Komidjan area of central Iran.

## 9.2 THE REGIONAL CONTEXT

Iran is an interesting site for a sustainability assessment because of the many obstacles it faces in achieving sustainability in its agricultural systems. Iran faces difficulties in at least two of the four types of factors disrupting agricultural sustainability — biophysical, socioeconomic, technical, and institutional (Farshad, 1997).

Biophysically, Iran is situated in one of the agriculturally unfavorable parts of the word (i.e., too cold, too dry, too hot, and/or too high in altitude) where it is very difficult to increase agricultural production without external capital input. Socioeconomically, high levels of poverty tend to encourage practices that increase production in the short term but undermine sustainability in the long-term.

Water scarcity makes irrigation  soil degradation (compaction, salinization, and waterlogging), water quality deterioration, vegetation depletion through overgrazing and/or drought, and land use competition resulting from urbanization affect the sustainability of agricultural systems.

During the last several decades, Iran's agricultural sector has been subjected to drastic changes and instability because of socioeconomic and technological upheaval. While many traditional social norms are preserved, new technology dictates changes that farmers may not accept. In this context, the semi-arid agricultural areas of Iran are especially vulnerable because of dry climate, salt affected and/or excessively calcareous soils, low soil organic carbon content, shortage of surface water, overexploitation of groundwater with drastic lowering of the water table depth, population growth, and inappropriate changes in land tenure.

### 9.2.1  Biophysical Conditions

The semi-arid regions of Iran are characterized by alternating warm and cold seasons. Variations in temperature are considerable, with a mean maximum monthly temperature of 30.0°C in summer and a mean minimum monthly temperature of 5.0°C in winter. Day and night temperatures are also strongly contrasting. The monthly precipitation exceeds the potential evapotranspiration in only 7 months of the year. These regions mainly belong to the bioclimatic zones termed "thermomediterranean" and "mesomediterranean" (xeric index of 40 to 150), but some fall within the "xerothermomediterranean" zone (xeric index of 150 to 200) and the "cold steppic" zone (a dry and freezing period of 5 to 8 months). The xeric index is based on the Gaussen method and defined as the number of biologically dry days (Sabeti, 1969).

Large areas in the Alborz and Zagros mountains, stretching along the northern and western borders, respectively, have a semi-arid climate. Semi-arid conditions occur in the mountainous areas, including hills, ridges and intermontane basins or valleys ranging in elevation from 1000 to 2000 m. Agricultural activities mainly concentrate in the intermontane valleys; steep slopes in the mountains are used for rangeland. Semi-arid conditions permit dry farming, at least during one season, in contrast with arid regions where dry farming is impossible (Farshad, 1990).

The Hamadan-Komidjan area, in the Hamadan province of western Iran, properly represents the semi-arid conditions typical of Iran. The provincial capital Hamadan, an ancient town at the skirt of the Alvand mountain in the central Zagros mountain ranges, is situated about 400 km southwest of Tehran (see Figure 9.1).

Except for the Alvand mountain, which is formed of granitic and metamorphic rocks, the rest of the area is composed of limestone, sandstone, and shale. The main landforms are mountains, hills, piedmonts, and the Gharachai and Sharra valleys. All rivers originate from the Alvand mountain, except the Sharra river, which has its catchment in the Shazand mountains. The rivers have a seasonal rhythm, with the highest discharge from March to April and the lowest from June to August. Quaternary sediments, occupying a large part of the study area as piedmont glacis and fans, play a significant role in the groundwater recharge. Most deep wells are located in the piedmont, and range from 40 m to more than 100 m in depth. The dominant soils are calcixerollic, typic, and fluventic xerochrepts. Other soils are petrocalcic xerochrepts, typic and lithic xerorthents, natrixeralfs, and salids. Salt- and sodium-affected soils occur in the eastern part of the study area, mainly in the Sharra valley.

## 9.2.2 Agricultural Systems

In the Hamadan-Komidjan area, traditional and modern agriculture is practiced although traditional farming is steadily disappearing. Traditional farming includes the use of animal drawn wooden ploughs, local seeds, *ghanat* (underground tunnels), *cheshmeh* (springs) and/or harvested runoff water, and the absence of agricultural machinery and chemicals (see Figure 9.2).

A traditional production unit is a complex system of interrelated activities carried out by a household. It includes three main components: crop farming, animal husbandry, and handicraft production. Functional integration and temporal distribution of the activities make it necessary for all family members to participate full-time throughout the year. Oxen, cows, sheep, goats, hens, and pigeons are common. Milk products, eggs, meat, flour from wheat and barley, vegetables, fruits, leather, and wool are produced. The large variety of products generated help mitigate risks from climatic (e.g., drought) to economic (e.g., fluctuations in the world market price).

In contrast, modern farming systems are characterized by the use of water emanating from deep wells and the Yalfan dam, improved seeds, machinery (at least tractors), chemical fertilizers, herbicides, and pesticides (see Figure 9.3). The introduction of new sources of energy, technology, and machinery has changed the relationship between inputs and outputs in the traditional production system. Crop production, animal husbandry, and rural industries are no longer interdependent activities.

**Figure 9.1** The Hamadan province, Iran.

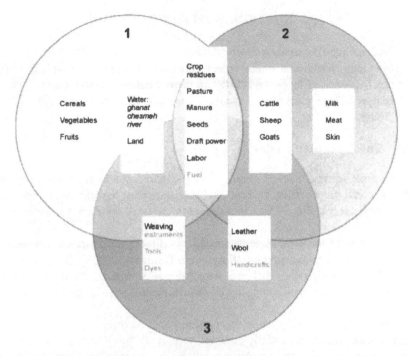

**Figure 9.2**    Model of a traditional agricultural system. This system is based on the integration of three interdependent production sectors within one household unit: (1) cultivation, (2) animal husbandry, and (3) rural crafts. Production is oriented toward family consumption; surpluses are exchanged among households in the same village.

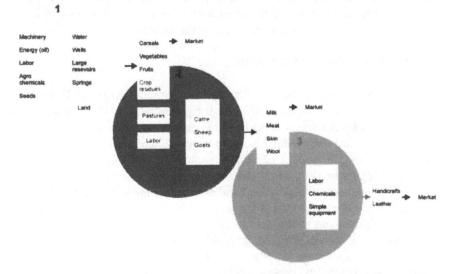

**Figure 9.3**    Model of a modern agricultural system. This system is based on three independent production sectors belonging to separate household units: (1) cultivation, (2) animal husbandry, and (3) rural industry. Production is market oriented and each sector specializes in delivering intermediate and final products.

## 9.3 SUSTAINABILITY ASSESSMENT

The sustainability of two irrigated wheat land-use systems in the Hamadan-Komidjan area, one under traditional management and another under modern management, were each assessed using two methods: an energy balance analysis and a socioecologic analysis. These assessments used a conceptual reference system called the Six-Pillar Model.

### 9.3.1   The Six-Pillar Model

A sustainable system has six requirements: environmental soundness, economic viability, social acceptability, institutional manageability, agrotechnical adaptability, and political acceptability. These requirements can be considered "pillars" on which a sustainable system is built.

Since none of the requirements (pillars) is directly measurable, relevant indicators are required to assess them (Smyth and Dumanski, 1993). Because the same indicators are often used in different ways to assess more than one pillar, a three-level model was designed, made up of requirements, criteria, and indicators (see Table 9.1).

Assessing sustainability using this model would require a large team of experts, therefore assessments are usually confined to parts within one or two of the pillars. Depending on the objective, emphasis might be put on economic, sociologic, and/or environmental aspects. In some cases, especially when economic constraints are involved, natural resources are either disregarded or only marginally taken into account (Ikerd, 1990; Norgaard, 1975, 1984). Even when dealing with only one of the pillars in the model (e.g., environmental soundness), many data from different sources are required to satisfy the criteria and indicators; rules are also needed to take care of all possible interactions among the indicators.

### 9.3.2   Energy Balance Analysis

Agroecosystems depend on both ecologic and agricultural forms of energy. The ecologic energy includes solar radiation for photosynthesis and appropriate atmospheric conditions, while the agricultural energy includes biologic (e.g., labor, manure application) and industrial components. When a natural system capable of producing a certain amount of energy containing biomass is converted into an agroecologic system, the natural capability limit is often exceeded by adding energy inputs. The greater the input of external energy, the more the natural capability of the system can be exceeded, and the less sustainable the system becomes. Because of this relationship, an analysis of an agroecosystem's input/output energy balance ratio can be a comprehensive indicator of its sustainability.

Since energy use data are often difficult to obtain or lack accuracy, our energy balance analysis required cross checking through multiple interviews and direct *in situ* measurements, such as crop cutting in a farmer's field for yield estimation.

Modern farming in Iran is based on a set of highly mechanized operations, which consume large amounts of energy in terms of labor and use of machinery (Koocheki

Table 9.1  Requirements, Criteria, and Indicators Used to Measure Agricultural Sustainability

| Requirements (pillars) | Criteria | Indicators[a] |
|---|---|---|
| 1. Political acceptability | Ease of employment<br>Government willingness<br>Life expectancy | Political attractiveness of the system<br>Working age<br>Birth rate (3,4) |
| 2. Economic viability | Attractiveness of land to non-agricultural users<br>Food self-sufficiency<br>Efficiency of inputs<br>Meeting market requirements<br>Net-farm profitability | Distance to non-agricultural area<br>Net present value of land<br>Average income/family<br>Imports as a percent of merchantable exports (3)<br>Working population % (3)<br>Potential/actual working population (3)<br>Surface area of cultivated land (3,5)<br>Yield/ha (5,6) |
| 3. Institutional manageability | Favorability of age distribution<br>Labor availability<br>Migration balance<br>Security of water supply | Average age (4)<br>Migration rate/year (4)<br>Population/land ratio (5)<br>Birth rate (1,4)<br>Imports as a percent of merchantable exports (2)<br>Working population % (2)<br>Potential/actual working population (2)<br>Surface area of cultivated land (2,5) |
| 4. Social acceptability | Human health<br>Infant mortality<br>Labor availability<br>Degree of welfareness<br>Literacy rate | Subsidy status<br>Mortality rate/year<br>Infant mortality rate/year<br>Literate/illiterate ratio<br>Birth rate (1,3)<br>Number of physicians in the region<br>Average age (3)<br>Migration rate/year (3) |
| 5. Agrotechnical adaptability | Access to groundwater<br>Agricultural production density<br>Attractiveness of land to non-agricultural users<br>Weed control<br>Pest control<br>Irrigation system status<br>Tillage | Methods of weed control<br>Methods of pest control<br>Surface area of cultivated land (2,3)<br>Yield/ha (2,6)<br>Tillage method (6)<br>Present observed erosion (6)<br>Precipitation (6)<br>Groundwater depth (6)<br>Potential water recharge (6)<br>Irrigation efficiency (%) (6)<br>Manure applied (6)<br>Mode of water supply (6)<br>SAR of water (6)<br>Water discharge (6)<br>Change of watertable depth (6)<br>Population/land ratio (3) |

continued

**Table 9.1 (continued)  Requirements, Criteria, and Indicators Used to Measure Agricultural Sustainability**

| Requirements (pillars) | Criteria | Indicators[a] | |
|---|---|---|---|
| 6. Environmental soundness | Soil alkalinity | Tillage method (5) | Soil pH |
| | Soil salinity | Present observed erosion (5) | Thickness of A horizon |
| | Soil compaction | Precipitation (5) | Bulk density |
| | Soil drainage condition | Groundwater depth (5) | Soil consistency |
| | Soil erosion status | Potential water recharge (5) | EC of soil |
| | Deterioration of topsoil structure | Irrigation efficiency (%) (5) | Drainage class |
| | Root penetration in soil | Manure applied (5) | ESP of soil |
| | Soil water holding capacity | Mode of water supply (5) | Gypsum content |
| | Biological activity in soil | SAR of water (5) | Water infiltration rate |
| | Water quality | Water discharge (5) | CaCO₃ content |
| | Water sufficiency | Change of watertable depth (5) | Moisture content (of soil) |
| | Influence of agricultural system on soil | Water salinity | Organic matter content of topsoil |
| | Influence of agricultural system on water | Soil structure | Yield/ha (2,5) |
| | | Topsoil texture | |
| | Influence of agricultural system on air | Subsoil texture | |
| | Attractiveness of land to non-agricultural users | | |

[a] The number (1–6) assigned to an indicator identifies the other requirements with which it is associated.

From Farshad, A., ITC publication 57, 1997.

and Hosseini, 1990). Land preparation starts with plowing in Spring, followed by leveling using an implement called a *mauleh*. Sowing takes place in the last week of October, using a deep row crop cultivator (*amigh kar*). Crop care includes fertilizer application, spraying of herbicides, and irrigation. Two kinds of energy input are involved: direct energy, energy spent in plowing and irrigation, and indirect energy, such as energy embodied in seeds and fertilizers (see Tables 9.2, 9.3, and 9.4). The analysis shows that the consumed energy (41.841 + 10.464 = 52.304 Gj/ha) is approximately half of the energy produced (99.5 Gj/ha), which yields an input/output ratio of roughly 1 to 2.

Traditional wheat farming in Iran is based on a trial proven sequence of activities, including land preparation by plowing and leveling, sowing, application of irrigation and fertilizers, and harvest. A traditional wooden plow pulled by oxen plows the land three times. Plowing takes two days per hectare. Before the third plowing, the land is irrigated to reach field capacity, which takes six to seven days, and seeds are broadcasted. The amount of seed per hectare varies between 120 and 150 kg.

Table 9.2   Direct Energy Consumed by the Mechanized Wheat System

| Activity | Time (hr/ha) | Number of treatments | Fuel used (L/ha) | Energy value | Total required energy (Gj/ha) |
|---|---|---|---|---|---|
| Plowing | 5 | 2 | 40 | 42.7 Mj/L | 3.416 |
| Leveling | 1 | 1 | 10 | 42.7 Mj/L | 0.427 |
| Sowing | 1 | 1 | 15 | 42.7 Mj/L | 0.640 |
| Irrigation | 7 | 5–6 | 150 | 42.7 Mj/L | 35.227 |
| Harvest | 2 | — | 40 | 42.7 Mj/L | 1.708 |
| Transportation | — | — | 5 | 42.7 Mj/L | 0.213 |
| Labor | 110 | — | — | 1.9 Mj/hr | 0.210 |
| Total | 126.5 | — | 260 | — | 41.841 |

Table 9.3   Indirect Energy Consumed by the Mechanized Wheat System

| Activity | Amount (kg/ha) | Energy value | Total required energy (Gj/ha) |
|---|---|---|---|
| Nitrogen (N) | 34 | 75 Mj/kg | 2.550 |
| Phosphorus (P) | 48 | 13 Mj/kg | 0.624 |
| Insecticide | 1 | 180 Mj/kg | 0.180 |
| Seed | 250 | 18 Mj/kg | 4.500 |
| Machinery | 30 | 87 Mj/kg | 2.610 |
| Total | — | — | 10.464 |

Table 9.4   Energy Output of the Mechanized Wheat System

| Output | Yield (kg/ha) | Energy value | Energy output (Gj/ha) |
|---|---|---|---|
| Wheat (grain) | 3750 | 14 Mj/kg | 52.5 |
| Straw | 4700 | 10 Mj/kg | 47.0 |
| Total | — | — | 99.5 |

Leveling follows using a simple wooden lath (*mauleh*) pulled by two oxen. Additional manual leveling might be necessary, especially in the corners of the field not reached by the *mauleh*.

The land is irrigated three times. Harvested runoff water (*seilaub*) from the mountains is the water most used. The irrigation interval is 12 days. Urea is applied in March at the rate of one bag per hectare, an operation that takes eight hours. Manure is still applied by some farmers instead of urea. It is mixed with the soil while plowing the land. The manure produced by 10 cows is said to be enough for three hectares of land. In the last week of July the crop is harvested, thrashed, winnowed for grain straw separation, and sieved. All activities are carried out by hand or with animals. It requires two 12 hour days to harvest one hectare. Thrashing the production takes five days per hectare, although some farmers use machines for this task.

The analysis shows that traditional agriculture consumes little energy (6.061 Gj/ha), while producing a large amount of energy (46.838 Gj/ha). This equals an input/output ratio of 1 to 8, much better than the 1 to 2 ratio of the mechanized system (see Tables 9.5 and 9.6). If it is assumed that the 1:8 ratio of the traditional system represents the threshold of sustainability in this region, then the mechanized system approaches the realm of unsustainability. However, the latter produces twice as much wheat as the former and is thus better able to satisfy, at least in the short term, the growing market demand.

### 9.3.3  Socioecologic Analysis

Energy flow might be the basis on which economists and environmentalists examine an agricultural system, but it addresses only a limited number of the criteria in the

Table 9.5   Energy Input of the Traditional Wheat System

| Input | Energy value | Amount/ha | Total required energy (Gj/ha) |
|---|---|---|---|
| Labor | 2.10 Mj/hr | 330 hours | 0.69 |
| Oxen | 2.9 Mj/hr | 190 hours | 0.56 |
| Machinery | 0.4 Mj/L | 60 L gas-oil | 0.024 |
| Fertilizer | 60 Mj/kg | 50 kg | 2.99 |
| Manure | 1 kj/kg | 1600 kg | 0.002 |
| Seed | 14 Mj/kg | 130 kg | 1.795 |
| Total | — | — | 6.061 |

Table 9.6   Energy Output of the Traditional Wheat System

| Output | Energy value | Amount/ha | Energy output (Gj/ha) |
|---|---|---|---|
| Grain (wheat) | 14 Mj/kg | 2000 kg | 28.438 |
| Straw | 9 Mj/kg | 2000 kg | 18.400 |
| Total | — | — | 46.838 |

Six-Pillar Model. Another approach to assessing sustainability, which secures some transversality through the pillars, is the socioecologic analysis.

Under natural conditions most land uses are sustainable (Stewart et al., 1990). In the past when the world was less populated, land was more commonly used in a friendlier way, respecting fallow periods and other traditional soil and water management practices. Time was available to allow disturbed or depleted agroecosystems to recover. Over a period of time, an equilibrium was reached between natural processes and human practices. The population growth and the changes in the social structure that have accompanied modernization, however, have to disrupted this co-evolutionary equilibrium.

In a socioecologic analysis, the traditional and modern agricultural systems are compared in their relationships to natural and human resources (see Table 9.7). In the area of natural resources, the ways the two systems cope with climatic risks, water scarcity, and soil restrictions are contrasted. Modern agricultural management tends to overcome these limitations by applying hard technology (e.g., deep wells and heavy machinery, which lead to ground water depletion and land degradation, respectively). In contrast, traditional land management uses local knowledge tested over the centuries for sound water and land management.

Both systems are also compared in how they use human resources and satisfy human needs. Endogenous factors, including farmers' knowledge, access to resources, and production and consumption objectives, as well as exogenous factors such as social organization, institutional support, and population dynamics, are assessed. In general, the development of modern agriculture is based on technological, social, economic, and institutional requirements that create new production conditions incompatible with the structure and functioning of the traditional communities.

## 9.4 COMPARISON OF THE ASSESSMENT METHODS

The two assessment methods were compared using the matrix in Table 9.8, which is organized according to the requirements and criteria of the Six-Pillar Model. Each method's contribution is indicated in the two right-hand columns.

The socioecologic analysis contributes to the assessment of many criteria, mainly qualitative, which provides a comprehensive picture of the agricultural systems. Comparatively, the energy balance analysis uses fewer criteria but constitutes an attractive approach because it generates quantitative results. The socioecologic approach is particularly appropriate for assessing criteria that describe the environmental soundness and social acceptability of agricultural systems, while the energy balance approach successfully handles criteria referring to economic viability and agrotechnical adaptability. The two methods can therefore be seen as complementary, providing a complete picture of a system's sustainability.

**Table 9.7  Socioecologic Analysis of the Agricultural Systems in the Hamadan-Komidjan Area**

| | Factors | Traditional Systems | Emerging Systems (In development) |
|---|---|---|---|
| Natural resources | Climate | Semi-arid type of climate; unreliable precipitation | Same climatic constraint |
| | Water | Ghanats, springs and rivers as water sources for irrigation | (Semi) deep wells; ghanats and springs drying out because of intensive groundwater exploitation |
| | Soil | Sustainable land management based on local knowledge about soil behavior | Introduction of heavy equipment and modern technology, often leading to soil degradation |
| Human resources; endogenous factors | Farmers' knowledge | Skilled workers with experience in carrying out the diverse activities of an integrated farm unit | Farmers cannot easily cope with changes |
| | Access to resources | The required resources (material and human) are available and easily mobilized | Natural resources do not satisfy the needs of the growing population; human resources are not easily available because of changes in the social structure |
| | Production and consumption objectives | Production satisfies family consumption and urban demand; city people depend on villagers | Production is mainly market-oriented; villagers shop in towns |

| Human resources; exogenous factors | | | | |
|---|---|---|---|---|
| Social Organization | | Traditional values and norms | Fully respected | Imported western norms disrupt community life |
| | | Resource distribution mechanism | Boneh is the production unit; landlord and community regulate the use of land and water | Land reform causes the disappearance of landlords; farmers own the land, but community management is lacking |
| | Institutional support | Agricultural extension | Not required | Available, but not efficient |
| | | Technology | Simple homemade tools; remarkable water management | New technology is introduced, but without providing the necessary training to farmers |
| | | Credit | Communal way of life, where credit has no meaning | Available, but often banks are too business oriented and farmers are not used to seeking credit |
| | | Health | Natural rural life style; absence of official medical care, welfare support and birth control | Improving medical and welfare conditions have brought about a large population growth |
| | Population dynamics | Spontaneous migration | Very seldom | Very common, sometimes to the extent of breaking family bonds |
| | | Organized migration | Social structure does not permit it | Common seasonal migrations: groups of farmers specialized in the cultivation of vegetables move to places with sufficient water provision, sometimes over large distances |

From Farshad, A. and Zinck, J.A., *Ann. Arid Zones*, 34(4), 1995.

Table 9.8   Comparison of the Two Assessment Methods

| Requirements (pillars) | Criteria | Energy Balance Analysis | Socio- Ecologic Analysis |
|---|---|---|---|
| 1. Political acceptability | Ease of employment | — | — |
| | Government willingness | — | –/+ |
| | Life expectancy | — | –/+ |
| 2. Economic viability | Attractiveness of land to non-agric. users | — | — |
| | Food self-sufficiency | — | –/+ |
| | Efficiency of inputs | ++ | + |
| | Meeting market requirements | + | + |
| | Net-farm profitability | ++ | ++ |
| 3. Institutional manageability | Favorability of age distribution | — | — |
| | Labor availability | — | + |
| | Migration balance | — | ++ |
| | Security of water supply | — | ++ |
| 4. Social acceptability | Human health | — | ++ |
| | Infant mortality rate | — | –/+ |
| | Labor availability | — | ++ |
| | Degree of welfareness | + | + |
| | Literacy rate | — | + |
| 5. Agrotechnical adaptability | Access to groundwater | — | + |
| | Agricultural production density | ++ | + |
| | Attractiveness of land to non-agric. users | — | — |
| | Weed control | — | — |
| | Pest control | + | + |
| | Irrigation system status | –/+ | –/+ |
| | Tillage | –/+ | + |
| 6. Environmental soundness | Soil alkalinity | — | ++ |
| | Soil salinity | — | ++ |
| | Soil compaction | — | ++ |
| | Soil drainage condition | — | ++ |
| | Soil erosion status | — | ++ |
| | Deterioration of topsoil structure | — | ++ |
| | Root penetration in soil | — | ++ |
| | Soil water holding capacity | — | ++ |
| | Biological activity in soil | –/+ | ++ |
| | Water quality | — | ++ |
| | Water sufficiency | –/+ | ++ |
| | Influence of agricultural system on soil | –/+ | ++ |
| | Influence of agricultural system on water | –/+ | ++ |
| | Influence of agricultural system on air | –/+ | ++ |
| | Attractiveness of land to non-agric. users | — | + |

— = no contribution;  –/+ = indirect contribution;  + = slight contribution;  ++ = strong contribution.

# REFERENCES

Farshad, A., A generalized overview of the effect of agriculture systems on the quality of soil in the semi-arid regions of Iran, 14th ISSS Congress, Kyoto, Japan. 6, 219–220, 1990.

Farshad, A., Analysis of integrated soil and water management practices within different agricultural systems under semi-arid conditions of Iran and evaluation of their sustainability, ITC, Publication 57, Enschede, The Netherlands, 1997.

Farshad, A. and Zinck, J.A. , Seeking agricultural sustainability, *Agric. Ecosystems Environ.*, 47, 1–12, 1993.

Farshad, A. and Zinck, J.A., The fate of agriculture in the semi-arid regions of Iran: A case study of the Hamadan region, *Ann. Arid Zones*, 34(4), 235–242, 1995.

Ikerd, I.J., Agriculture's search for sustainability and profitability, *J. Soil Water Conservation,* Jan/Feb 18–23, 1990.

Koocheki, A. and Hosseini, M., *Energy Flow in Agricultural Ecosystems* (in Persian), Entesharat-e-Djavid, Mashhad, Iran, 1990.

Norgaard, R.B., Scarcity and growth: how does it look today? *Amer. J. Agric. Econ.*, 57(5), 810–814, 1975.

Norgaard, R.B., Coevolutionary development potential, *Land Econ.*, 60(2), 160–173, 1984.

Sabeti, H., Les études bioclimatiques de l' Iran. University of Tehran, publ. 1231, 1969.

Smyth, A.J. and Dumanski, J., FESLM: an international framework for evaluating sustainable land management, *World Soil Resources Rep.*, 73, FAO, Rome, 1993.

Stewart, B.A., Lal, R., El-Swaify, S.A. and Eswaran, H., Sustaining the soil resource base of an expanding world agriculture, 14th ISSS Congress, Kyoto, Japan. 7, 296–301, 1990.

Zinck, J.A. and Farshad, A., Issues of sustainability, *Can. J. Soil Sci.*, 75, 407–412, 1995.

# CHAPTER 10

# Coevolutionary Agroecology: A Policy Oriented Analysis of Socioenvironmental Dynamics, with Special Reference to Forest Margins in North Lampung, Indonesia

Remi Gauthier and Graham Woodgate

## CONTENTS

1-8493-0894-1/01/$0.00+$.50
© 2001 by CRC Press LLC

## 10.1 INTRODUCTION AND OVERVIEW

This chapter highlights the coevolutionary character of agroecosystems and the socioenvironmental relations that drive and are driven by them. In the early sections a broad range of theoretical work is discussed to demonstrate a convergence of interest within both natural and social sciences around a number of important principles related to social and natural dynamics: the importance of context; the duality of structure; and the unpredictable nature of change. These sections are followed by a summary of a detailed empirical study of coevolutionary processes as experienced within two ethnically distinct but geographically contiguous rural communities in the province of Lampung in southern Sumatra, Indonesia. The chapter concludes by drawing out some important lessons for sustainable rural development policy.

## 10.2 THE INTERDISCIPLINARY IMPERATIVE
## AND ITS INSTITUTIONAL CONSTRAINTS

Systems of agricultural production are simultaneously economic, political, cultural, historical, ecological, agronomic, and environmental, and thus it is no surprise that the field of agroecology has developed as a multidisciplinary endeavor including all these areas of study. Research springing from particular traditional disciplines generates an overall understanding of agroecological processes that is incomplete. If we want to move agriculture in a more sustainable direction, we must develop a more complete understanding of agricultural production systems; this requires a multidisciplinary approach to agroecological research.

Many of us engaged in efforts to model agroecological processes and develop indicators of sustainability are using inter- or transdisciplinary approaches. Gliessman, for example, indicates the need to do so when he says that, "For any agroecosystem to be sustainable a broad series of interacting ecological, economic, and social factors and processes must be taken into account" (1990). An understanding of processes at the level of the ecosystem, suggests Gliessman, should interface with the "even more complex aspects of social, economic and political systems within which the agroecosystems function." The value of interdisciplinarity for moving towards sustainability is clearly identified in Conway's claim that the "critical dynamics of agroecosystems arise precisely where the socio-economic processes interact with the ecological" (1990). As Gliessman notes:

> The challenge for agroecology is to ... find a research approach that consciously reflects the nature of [productive activities] as the coevolution between culture and environment, both in the past and the present. The concept of the agroecosystem can (and should) be expanded, restricted, or altered as a response to the dynamic relationships between human cultures and their physical, biological, and social environments (1990:8).

While the need for interdisciplinarity in agroecosystem analysis may not be particularly contentious for those engaged in the field, Conway (1990) suggests that the matter often receives little more than lip service. Some of the reasons why this might be so will be immediately obvious to anyone who has attempted to make progress in this direction. Despite much fine rhetoric on the part of research funding bodies, resources for novel research that attempt to develop interdisciplinarity are difficult to secure; disciplinary boundaries and deeply held epistemological canons are still vigorously defended by academic journals and research assessment panels.

In the real world, beyond the ivory towers of academia we find other reasons for a lack of commitment to interdisciplinary approaches to sustainability. These reside in the divisions of bureaucratic governments and the differing imperatives (from economic growth, through social welfare, to nature conservation) followed in devising policy interventions. They are important vested economic interests which maintain that improved social welfare and environmental protection can only be tackled successfully within a market economy unfettered by restrictive policy. The central argument of these interest groups is simply that many ecological and social variables that are so important to sustainability are not subject to property rights and, thus, have yet to be utilized. If prices were attached to these variables, so the argument goes, the market would ensure their efficient distribution (Pearce et al., 1989).

The aim of this chapter is not to discuss such claims in detail; the purpose of referring to these debates is to highlight the dissonance between the modern world order and the goal of sustainability, which is essentially a postmodern concept. To address sustainability, a holistic transdisciplinary paradigm is essential. The prevalent view based on reductionist science impedes the search for sustainability by separating social, cultural, economic, and natural dimensions. This is a result of modernism, which for centuries has shaped the way the world is perceived in the West.

The birth of the modern era is linked by many to the scientific revolution of the 17th century and is characterized by the notion of progress and development toward a future in which people, through science and technology, would be able to domesticate and control nature (Clark, 1993). The central philosophical framework of the modern era is mechanistic positivism. Nature is viewed as some kind of giant clockwork machine, the workings of which are only amenable to rigorous scientific investigation. According to this logic, successful control of nature is simply a matter of generating sufficient understanding, achieved by breaking the machine down into its constituent parts (reductionism), before rebuilding it in order to realize the desired objectives. The validity of this approach rests on the Aristotelian/Cartesian separation of mind and matter, and the notion that in the process of investigation there is no impact of the researcher on the researched or of the researched on the researcher. This allows positivists to claim that scientific knowledge is objective knowledge, based on evidence derived from empirical data, replicated, and verified through scientific experiments. It is a system of thought and action that generates specialist knowledge concerning specific elements of nature.

The postmodern worldview is quite distinct, and, in terms of what we might tentatively call postmodern science*, stresses the reflexivity or self-awareness of the scientist, the absence of universal, objective truths, and a desire to cross traditional disciplinary boundaries. Agroecology is clearly allied to a generalized environmental

---

* The reader might like to compare the idea of "post-modern science" with Funtowicz and Ravetz's (1993) notion of a "post-normal science."

movement in society. Environmentalism is a good example of the new social movements (organized around issues other than class) that are considered some of the
defining features of postmodern society. Gandy's 1997 article on the links between
environmentalism and postmodernism concludes that:

> The most important lesson to emerge from any serious engagement between post-
> modernism and environmentalism is that we cannot understand changing relations
> between society and nature by relying on ahistorical and positivist modes of expla-
> nation which refuse to engage with the social and ideological dimensions of envi-
> ronmental discourse. The agenda for environmental research has suddenly become
> far more complex and interdisciplinary than has hitherto been the case. This places
> a major intellectual burden on environmental research to provide explanations for
> environmental degradation that are capable of contributing to policy discourse without
> presenting partial and misleading accounts of environmental change.

## 10.3 MOVEMENTS TOWARD INTERDISCIPLINARITY

Despite the situation depicted in the previous section, there are cases of well-
formulated interdisciplinary research receiving critical acclaim. These works come
from a variety of disciplines but share a number of important concepts and principles.

Our own model of interdisciplinarity is based on the attempts of Marx and Engels,
in the middle of the 19th century, to erase the distinctions between natural and social
science. These distinctions, they believed, were inadequate in situations where the
natural is increasingly affected by the human. As Engels (1959) wrote:

> Let us not ... flatter ourselves overmuch on account of our human victories over
> nature. For each such victory takes its revenge on us. Each victory, it is true, in the
> first place brings about the results we expected, but in the second and third places it
> has quite different, unforeseen effects which only too often cancel the first. ... Thus
> at every step we are reminded that we by no means rule over nature like a conqueror
> over a foreign people, like someone standing outside nature but that we ... belong to
> nature, and exist in its midst.

Eventually, suggested Marx (1975), a single science would have to be created:
"The idea of one basis for life and another for science is from the outset a lie....
Natural science will in time subsume the science of man just as the science of man
will subsume natural science: there will be one science."

Dickens (1997) points out that Engels, in *The Dialectics of Nature* (1959),
attempted to map out the "one science" framework suggested by Marx. Engels'
model suggested that while physics and chemistry are appropriate for explaining the
material world, the emergence of life brings with it its own distinctive set of organizational principles, dynamics, and driving forces. Thus, while physics and chemistry can tell us something about the biological world to which they give rise, we
require additional insights from biology to render the living world intelligible.
Similarly, physics, chemistry, and biology can contribute to our understanding of
society, but the appearance of Homo sapiens on the scene introduces yet another set

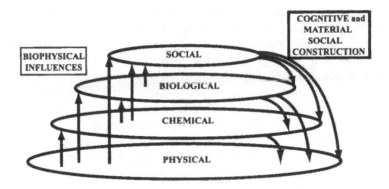

**Figure 10.1** Socioenvironmental relationships and dynamics.

of organizational principles, dynamics, and driving forces. The one science model, then, is a model of emergent properties, taking at its core the fact of change.

There is another important social dimension to this model. While the idea described thus far clearly suggests that nature makes society — that social life arises from, and is strongly conditioned by, biophysical processes* — we also need to recognize the idea that society makes nature. The key premise of constructivist sociology is that it is the way in which we think about the world, how we construct it socially rather than our direct experience of it, which determines how we behave toward the world and each other.

In our model of socioenvironmental relations (Figure 10.1), while we indicate the influence of the physical, chemical, and biological on the social we also depict the impacts of the social on the biological, chemical, and physical. The implication of this model is that in changing the world, both physically and in terms of how we think about it, we also change ourselves.

Having sketched our own understanding of the basic relationships between the physical, biological, and social dimensions of agroecosystems, we can now consider a variety of initiatives in interdisciplinary socioenvironmental research, and discuss concepts that build upon the central principle of the indivisibility of society and nature.

In recent years the need to adopt alternative and more integrated analytical frameworks has challenged more academics from a range of disciplines. In ecology, the basic reductionist perspective of traditional science has been attacked as "inappropriate for understanding the emergence and evolution of living systems," (Allen, 1994) and research into nonequilibrium dynamics (McIntosh, 1987; Sprugel, 1991; Pahl-Wostl, 1995; Fiedler et al., 1997) has prompted the development of what some have called "chaotic ecology" (Allen, 1994), in which evolution is understood as a nonlinear, and thus inherently unpredictable, process.

According to chaotic models, nature throws up a multiplicity of variations in both the physical and the biological elements of ecosystems. When complex,

---

* It is worth noting here that the idea that social life is so strongly conditioned by biophysical factors is a complete anathema to many social scientists. This stems from Durkheim's famous claim that social phenomena can only be understood through recourse to "social facts."

nonlinear systems are modeled mathematically their structures only change at certain moments in time. Evolution is characterized by phases of apparent stability and rapid change. This suggests that change in biological systems occurs as a result of new or nonaverage patterns of behavior encountering some form of positive feedback. Paradoxically when conventional, mechanical, and linear models are used to predict the future, the very factors that are important in creating that future, the variations around the norm, are ignored. The nonlinear and chaotic nature of evolution, according to Allen (1994), means that the "organizing principle that underlies sustainable systems is the presence, the maintenance, and the production of microscopic diversity in the system! ... Ecological structure results from the working of the evolutionary process, and this in turn results from the nature of ecological structure."

This explanation of the relationship between ecological structure and evolution seems to echo the sociologist Anthony Giddens' understanding of the link between social structures and change. Social structures, writes Giddens (1979), "are both the medium and the outcome of the practices that constitute the system." In this respect, the similarities between natural and human systems appear at least as important as the differences.

Giddens' concept of "structuration," the means by which systems' participants reproduce or refashion social structures (1984), can be further illuminated by the concept of "possibility space," which Allen uses to explain the ecological structuring of human activity. Possibility space represents a multidimensional physical and social space that provides potential for new options and technologies to arise. New systems properties emerge when human activity is influenced by fresh information concerning the behavior of others and the nature of environments.

Allen (1994) writes that, "[I]n the real world, competitors, allies, clients, technologies, raw materials, costs, and skills all change. Any group or firm that fixed its behavior would sooner or later be eliminated, having no adaptive or learning capacity with which to respond." Thus, the structures of human societies, like the structures of ecosystems, are best understood as a "temporary balance between exploration and constraint." Allen's ecological understanding of structure relates to Giddens' (1979) assertion that social structures both enable and constrain people's intentional activities.

These ideas (that change is chaotic and structures both enable and constrain behavior) give rise to a third principle of socioenvironmental relations: they are heterogeneous across time-space. All three ideas are relevant to agroecology and have been used by ecologists, human geographers, and scholars of development studies. Another important element in better understanding relationships between society and nature, linked to the notion of time-space heterogeneity, is the importance of taking an historical, context specific perspective. Without it, many unidisciplinary studies have mistakenly implicated people in processes that are largely independent of human activity or viewed as natural and ubiquitous conditions largely of anthropogenic origin. As Meyer (1996) points out, "[T]he human imprint on the earth could be described as unmistakable, were it not often mistaken for the work of nature or natural phenomena for human imprints" (cited in Batterbury, Forsyth, and Thompson, 1997).

In recent years a number of publications have combined nonequilibrium ecology and poststructuralist sociology together with an historical perspective. Iconoclastic works such as Thomas and Middleton's (1994) *Desertification: Exploding the Myth*, Fairhead and Leach's (1996) *Misreading African Landscapes*, and Arnold and Dewees (1997) edited volume *Farms, Trees and Farmers*, have shown many examples of positivist generalization, concerning widespread environmental degradation in less industrialized countries to be well wide of the mark. These studies point toward the need for detailed, context specific, historically grounded empirical research, informed by an understanding of the heterogeneity of socioenvironmental systems and the nonlinear and unpredictable character of socioenvironmental change*.

The characteristics of socioenvironmental systems that we have so far outlined (their emergent properties, historical contingency, spatial heterogeneity, continual reformulation, inherent unpredictability, and the subjective way in which they are experienced by different social actors) are also brought together by Norgaard in his long standing work on coevolution (1984; 1994; 1997). Norgaard's work emphasizes how agricultural activities modify ecosystems and how ecosystem responses give cause for subsequent individual action and social organization.

The notion of coevolution is central to the case study that we shall present in the second half of this chapter.

## 10.4 COEVOLUTION BETWEEN SOCIETY AND NATURE

Norgaard's coevolutionary thesis is explored in detail in his 1994 book *Development Betrayed*. He explains how environmental factors affect the fitness of particular aspects of social systems while, at the same time, social systems influence the fitness of particular aspects of environmental systems. Norgaard divides social systems into knowledge, values, organization, and technology subsystems, each of which coevolves with the others and with environmental systems. All the systems change, whether by chance or design, and are affected by and effect change in the other systems. As the various components and features of each system put selective pressure on the components and features of the others, they all coevolve so that each reflects the others. As Norgaard (1997) notes, "Coevolution explains how everything appears to be tightly locked together, yet everything also appears to be changing."

In an earlier work, Norgaard (1984) emphasized how, during agricultural modernization, the social system frequently assumes the regulatory functions that were previously endogenous to the ecosystem or maintained by the individual farmer. He points out that in contrast to the classical view, which frequently attributes the high productivity of modern, capital intensive agriculture to technological mastery over nature, the "coevolutionary perspective emphasizes ... the

---

* A particularly good starting point for reviewing work of this genre can be found in a special number of *The Geographical Journal* (163(2), 1997) devoted to "Environmental Transformations in Developing Countries" — especially the introductory essay by Batterbury, Forsyth, and Thompson (pp. 126–132).

increasing task specialization and organizational complexity of maintaining feedback mechanisms between social actors and the environment."

The escalating complexity of social organization in modern industrial societies lengthens the chain of connections between society and nature so that the sustainability of highly industrialized agroecosystems becomes dependent not only on the maintenance of society-nature linkages but also on the upkeep of social relationships within complex actor networks. These relationships include links between the producers, individuals, and institutions (extension agents, credit banks, agriculture ministries, and development agencies, etc.) which impact the socioeconomic and policy environment in which productive activities are implemented. While at any given moment the current situation appears so intricate and complex as to be unchangeable, by standing back and taking an historical view of the situation, we can appreciate that the only constant element of the model is the fact that change, while proceeding at variable rates and in different directions, is continual. This suggests that sustainability needs to be understood as maintaining space for maneuver and adaptation in a continually changing world.

## 10.5 COEVOLUTION AND ENVIRONMENTAL TRANSFORMATION IN LAMPUNG

The research detailed in this part of the chapter attempts to illustrate coevolution through recourse to recent research among members of two ethnically distinct yet geographically contiguous rural communities in southern Sumatra (Gauthier, 1998). The Lampungese and Javanese farming communities were brought into close geographic proximity as a result of the government of Indonesia's population redistribution or transmigration programs. The research shows how the different histories of the two peoples produced different systems of values, knowledge, organization, and technology, and equally distinct agroecosystems. The impact of transmigration and agricultural development policies led to the emergence of new structures that enabled and/or constrained the livelihood strategies of families in the two communities. This initiated dynamic processes, tracing new pathways through "possibility space," giving rise to agroecological scenarios that are quite distinct from those envisaged by national agrarian development policy.

The lessons to be derived from the preceding theoretical discussion and the following case study will identify basic guidelines for more appropriate policies for facilitating rural livelihood and sustaining agroecosystem sustainability.

### 10.5.1 Biophysical and Sociocultural Overview of Lampung

The Province of Lampung is located between longitudes 105° 50' and 103° 40' east, and latitudes 3° 4' and 6° 45' south, at the southernmost tip of the island of Sumatra, Indonesia (see Figure 10.2). It is bounded on the north by the province of South Sumatra, to the south by the Sunda Strait, to the east by the Java Sea, and to the west by the Indian Ocean (BPS, 1994/1995).

Yusuf (1992) divides the province into five topographical types:

## Indonesia and the Province of Lampung

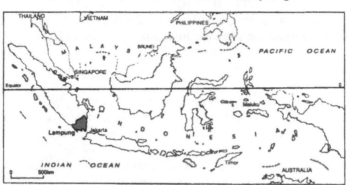

**Figure 10.2** Indonesia and the province of Lampung. (From Pain et al., 1989.)

- Hilly to mountainous, found mainly in the western part of the province, and represented by the southern section of the Bukit Barisan mountain range
- Undulating, with slopes between 8% and 15%, and elevations between 300m and 500m above sea level, which are widespread throughout the districts of South and Central Lampung, and heavily cultivated with both perennial and food crops
- Alluvial plains, spreading from the northern edge of Central Lampung district to the east and downstream from the major rivers such as Way Sekampung, Way Tulang Bawang, and Way Mesuji
- Tidal swamps, found along the east coast
- River basins, of which there are five major areas associated with the Way Tulang Bawang, Way Seputih, Way Sekampung, Way Semangka, and Way Jepara rivers

Rainfall in Lampung is abundant, with the majority of the province having between five and nine consecutive wet months (more than 200 mm rainfall/month) and three consecutive dry months (100 mm rainfall/month) per year. The main dry season occurs between July and August, although wet and dry seasons are not as clearly defined as in neighboring Java (Whitten et al., 1987). The temperature varies little throughout the year, with an average of 26 to 28°C at elevations of 30 to 60 m above sea level (BPS, 1994/1995).

Whitten et al. (1987) report that relatively fertile andosol and latosol soils occur mainly in the south of the province, with 2.4% of the area having andosol and 21.6% having latosol; less fertile, red-yellow podzolic soils cover 45.7% of the surface area of the province. The majority of the north of the province, where the current research was carried out, has red-yellow podzolic soils and hydromorphic alluvial soils (Levang, 1989). The former are well drained acid soils having thin organic and organic mineral horizons, susceptible to strong leaching and loss of fertility in high rainfall (Bridges, 1978), particularly where the forest canopy is lost. The latter soil type is poorly drained, creating swamp conditions (Bridges, 1978), and requires much preparation for agricultural use.

The population of the province has increased dramatically over the past thirty years; in 1961 it was estimated at 1,667,511, but by 1995 it had increased more than fourfold and reached 6,680,300 (Yusuf, 1992; BPS, 1994/95). This population growth is largely due to the Indonesian government's transmigration programs and the arrival of spontaneous migrants from other parts of Indonesia. The influx of people into Lampung has had environmental and social impacts. Population migration has generated environmental transformation processes, most visibly, forest clearance for agricultural production. Vast areas of the province have been cleared of natural vegetation and have been transformed into agriculturally productive land. Whitten et al. (1987) show only small areas of natural forest vegetation remaining in 1982, limited largely to the mountainous west of the province, the northern edge, and eastern coast.

Large scale migration has transformed the sociocultural make-up of the province. Indigenous Lampungese people are now a minority in their own land, accounting for only 20% of the total population of the province (BPS, 1994/95).

The ethnic Lampungese have distinct language and cultural traditions, which stand in marked contrast to those of the immigrant Javanese (Hadikusuma, 1989). These ethnic groups have traditionally practiced very different agricultures, with Javanese focusing on annual food crops, especially rice, and the Lampungese concentrating on perennial crops within intricate tree garden systems.

## 10.5.2 Research Processes and Methods

The research was conducted in three settlements in North Lampung (see Figure 10.3), chosen because of their ethnic compositions, their proximities to the forest, and their associations with governmental policies relating to transmigration and agricultural development. They were selected in order to investigate socioenvironmental change processes relating to the conversion of rain forest ecosystems to agroecosystems.

The methodology involved an intricate mixture of quantitative and qualitative methods, using an array of techniques ranging from participatory rural appraisal, structured questionnaire survey, and participant observation, to direct field observations and measurements. This variety of methods facilitated triangulation of data and increased the reliability of the information gathered.

In order to analyze the dynamic interaction of government policy, forest margin livelihoods, community structures, and environmental change, a hybrid coevolutionary framework similar to that outlined in the first part of this chapter was used. By placing social actors at the center of the analysis, it shed light on the ways in which these actors reproduce social and biophysical structure in their actions and challenge these structures through their agency. By recourse to actor-oriented analysis that is historically, socially, culturally, and ecologically embedded, the framework illuminated the role of actors, structure, and culture in environmental transformation and, in turn, shed light on how environmental change gives rise to sociocultural adjustments.

By using this approach to socioenvironmental research, biophysical factors can be seen as forces for change, as actors try to sustain their livelihoods by continual adjustment and readjustment of and to environmental fluxes. The coevolutionary

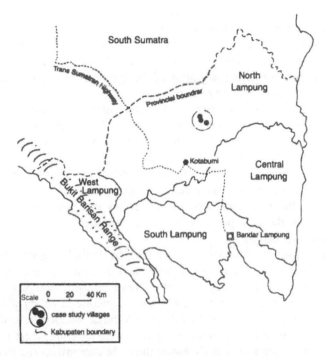

**Figure 10.3** Research sites in North Lampung. (Adapted from Sage, 1996.)

dynamics that are revealed are sensitive to change caused by both natural and social events. Examples of such events include the introduction of subsidies for agricultural inputs, shifts in livelihood strategy by a household, or the appearance of pests that affect the cropping system. The range, complexity, and interrelatedness of the factors affecting coevolutionary dynamics underscore the nonequilibrium, chaotic nature of socioenvironmental systems. These factors are exemplified by extracts from a detailed research project (Gauthier, 1998) that we have included in the following sections of this chapter.

## 10.5.3 The Structural and Historical Context of the Research Area

The social structure and policy milieux were examined in order to understand the structural setting in which the local actors have to perform. A longitudinal review of agricultural development in Indonesia provided an historical perspective that allowed the roots of present policies affecting forests and forest margin communities to be traced.

Various neoclassical and neopopulist agricultural development models have influenced policy formulation and have been instrumental in transforming Indonesia's natural environment into an agroenvironment. Government policies have affected the Indonesian economy, its development, and the environment, and have also set guidelines within which social actors interact and attempt to further their individual agendas.

In recent times a number of programs, such as transmigration and institutional strengthening, have been put in place to reinforce agricultural development policies. Transmigration is seen as a tool for national integration. It disperses Javanese farmers and increases the focus on food crop production throughout the archipelago, thus supporting the rice self-sufficiency aims of the government. Institutional strengthening aims at increasing implementation capacity.

A focus on green revolution rice production technology in agricultural development policy led to environmental consequences, including pest and disease proliferation and the transformation of vast tracts of forest into paddy fields. Long-term development plans envisage a shift from small to larger land holdings such as estates and plantations, and a general push towards modern agribusiness (Pemda Tk., I-Lampung, 1995), which will undoubtedly marginalize small farmers and increase the pressure on the remaining forests. The policy discourse of the government demonstrates how those in power view the environment — as a resource to be harnessed and transformed.

The effects of government policies have been particularly apparent in Lampung, which has been a recipient of agricultural development and transmigration for decades. The resulting demographic changes have brought about social and environmental transformations. The immigration and agricultural policy goals for Lampung force colonization of the forest for agricultural production, resulting in rapid landscape change. Farming communities at the forest edge are directly affected by the development policies on a daily basis, these, in turn, affect the livelihoods of local people.

The demographic changes that have occurred in Lampung over the past 60 years have had a great influence on the sometimes uneasy relationship between the Javanese and indigenous Lampungese populations. The influx of immigrants has also had a huge impact on the environment of the province in general and the forest in particular; directly, by reducing the area of forest by the opening up of land for immigrant agriculture (Pain et al., 1989), and indirectly, by driving the Lampungese into forested areas to establish new agricultural land of their own.

These demographic factors provide an explanation of environmental change when land use data are compared with demographic data at the provincial level. Detailed examination of the interactions between immigrant and indigenous communities and their local environments over time provides a more nuanced understanding of environmental transformation. The dynamics of this process were elucidated through an examination of the recent environmental history of the research area as told by its inhabitants.

In the 1960s the area was primary rainforest. It is only since then that intensive anthropogenic environmental change has occurred on a large scale. In 1970 six families established the first settlement followed by the establishment of two transmigration villages with some 700 families in 1984 (Pemda Tk., I-Lampung, 1983). The population increased dramatically with the arrival of immigrants; agricultural production shifted from extensive to intensive systems.

Over time, the landscape changed from closed canopy rainforest to mainly agricultural land with only a trace of forest remaining by the mid-1990s (Gauthier, 1998). This landscape change has impacted farmers and their livelihoods: As the forest receded, along with its vegetative canopy and predators, the environmental feedback functions of soil fertility maintenance and predator-prey relations — once internal to the forest ecosystem — had to be taken up by the local community.*

The social environment also changed. As official transmigrants and other spontaneous settlers established new villages and relations among communities, social space was created for land litigation. The resulting land tenure insecurity damaged the poorer farmers' livelihoods in two ways: it increased habitat and forage for rats on abandoned land; and, it discouraged farmers from taking a long-term approach to soil fertility conservation. A Javanese farmer, Bpk. Suliman, reports on a typical situation:

> I sharecrop _ hectare of sawah (wet rice), but most of the land around my plot isn't worked anymore — people have abandoned it because of the land disputes. I want to carry on cultivating rice, but with all the land around returning to scrub the rat problem is getting worse. This season I lost more than three-quarters of my crop to rats (Gauthier, 1998).

The local environmental history revealed by the area's inhabitants provides a timescape (Adam, 1997) that records the interactions between social actors, state policy, and the local environment. Changes in the environment have influenced changes in the behavior of social actors, further impacting on the environment and on other social factors, as individuals and communities resort to violence to resolve land disputes. This coevolutionary timescape illustrates the background and recent historical roots of present day practices, needs, and socioenvironmental relations.

## 10.5.4 Livelihoods and Environmental Use

The agroecological and livelihood strategies currently employed by households in both immigrant Javanese and local Lampungese communities are rooted in the structural and historical context of the research area. They represent both the outcome and the transcript of coevolutionary dynamics between the environment and the people.

The livelihood decision making process is one of optimization, based on the range of socioeconomic, cultural, and ecological factors with which social actors are faced. The actors seek to further their own agendas in a variety of ways. An example involves the use of agricultural packages from the government. Their components are used selectively within an array of activities that comprise local livelihoods. An example was observed when recipients of agricultural inputs associated with a smallholder sugar cane project diverted some of the fertilizer and biocide for use on their rice and other food crops (Gauthier, 1998). This provides evidence of the centrality of social actors within the coevolutionary framework, and also points to their ability to change the planned outcomes of rural development interventions.

The contrast between the predominantly food crop based agriculture of the Javanese farmers and the tree crop based systems of the Lampungese shows differential adaptation to local environmental conditions. While the food crop based

---

* Compare this experience with Norgaard's (1984) theoretical discussion of coevolutionary agricultural change.

Table 10.1 Average Percent Crop Loss Reported by Questionnaire
Respondents for Specific Crop and Pest Species

| Crop | Rats | Birds | Pigs | Monkeys |
|------|------|-------|------|---------|
| Wet Rice | 54.7 | < 1.0 | — | — |
| Upland Rice | 12.1 | 2.5 | 2.9 | — |
| Swamp Rice | 51.1 | < 1.0 | 3.7 | < 1.0 |
| Cassava | < 1.0 | — | 4.4 | 2.2 |
| Sugar Cane | < 1.0 | — | 1.0 | < 1.0 |
| Peanut | 1.0 | — | 2.1 | — |
| Maize | — | — | 3.5 | 3.6 |
| Lemon Grass | — | — | — | — |
| Vegetables | — | — | — | — |
| Soya Bean | — | — | — | — |
| Chili Peppers | — | — | — | — |
| Trees | — | — | — | < 1.0 |

Adapted from Gauthier (1998).

systems of the Javanese are plagued with pest and soil fertility problems, Lampung-ese tree gardens have minimal pest depredation and are better suited to soil conservation under local conditions. The situation with respect to crop depredation is illustrated in Table 10.1.

The degree to which agroecosystems are successful in achieving good financial returns while remaining suited to local environmental conditions influences the scope of households' livelihood strategies and, indirectly, household dynamics with regard to decision making possibilities and the division of labor. A comparison between Javanese and Lampungese households underlines the reflexive, negotiated nature of household decision making which, while remaining within cultural norms, stretches and transforms them. The decision making process is thus a mixture of pragmatic choice and culturally embedded tradition, that responds to a dynamic coevolutionary process between the environment and the social system.

Differences in agricultural and livelihood strategy between the Javanese and Lampungese communities are paralleled by differences in their attitudes toward environmental use. There are clear contrasts in attitudes toward food crops, tree crops, and forest between the groups, which fit within the two communities' respective social values and cultural norms. The contrast in attitudes toward agricultural crops tends to be bimodal. The Javanese consider themselves to be rice farmers. "Real farmers are rice farmers," according to Bpk. Sudiman a Javanese farmer (Gauthier, 1998). The Lampungese see themselves as tree farmers using systems based on traditional agroforestry and centered on rubber. Detailed discussions with farmers in the field demonstrated that Lampungese respondents saw their rubber agroforestry system as a means of protest against government policies that "Javanize" their way of life. This farming system helps to support their efforts to retain their distinct cultural identity.

There are also clear contrasts in attitudes toward the forest. The Javanese, on the whole, are reluctant to seek work in forest related activities. One Javanese farmer stated, "Only social outcasts or desperate people would work in the forest." (Gauthier, 1998). Off-farm work for Lampungese tends to be forest-based, with logging and bamboo harvesting the major off-farm activities. Some Javanese engage in forest related livelihoods, mainly eagle wood harvesting and hunting, but these individuals are marginal within their own communities and the hunters are often subjects of curiosity and derision. Nevertheless, the hunters are also seen as important information exchange agents, because in the course of their hunting expeditions they pass through different communities; this secondary activity helps to boost their social status within the community. A typical case of a farmer turned hunter is outlined below (Gauthier, 1998).

---

Bpk. Tukiman came to one of the case study settlements, Tegal Mukti, from Pringsewu in South Lampung district. In South Lampung he owned sawah and cultivated wet rice and some vegetables. In 1984 his land was gazetted as a conservation area and, along with many others, he was put on a local transmigration (Translok) program and moved to Tegal Mukti. On arrival he found that he had been allocated land far from the village, and that it was still forested. He planted upland rice in a clearing on this land, but the harvest was disastrous, with most of the crop eaten by rodents and trampled by wild pigs. He continued planting for two years with little success. During this time he started setting out some snares around his rice field in an attempt to control pests. He and his family ate the animals he caught . They are Hindus, originating from East Java.

The early years in Tegal Mukti were difficult for many of the new inhabitants, as heavy depredation by wild pigs continued. Bpk. Tukiman's acquired expertise in catching pigs became known within the village and some of his neighbors asked him to set snares on their land. Among the transmigrants were some nonMuslims who were prepared to acquire pig meat in exchange for money or favors, such as helping out on his land, but it was not until he was able to start selling meat in the village of Pulung, that has a significant Batak Christian community, that hunting became a viable livelihood option. On average, Bpk. Tukiman catches eight to ten pigs a month, providing his family an income which is relatively good by local standards. His son is now a hunter, too.

Their land is planted with sugar cane from a government program, but he has not yet made any money from cane cultivation. As soon as his loan is repaid, he would like to leave the program, but he is under pressure from the other members of his group to continue.

It is ironic that he is now constrained by having to repay loans to a government program aimed at alleviating poverty. Had he, like so many other farmers in Tegal Mukti, confined himself to cultivating food crops and sugar cane, he would be in no position to repay his loan, and would be forced, like the others, to work on the plantation. Instead, he has a livelihood that permits him to repay the loan and still meet daily needs.

He does not see hunting as a temporary or shameful occupation, stating with pride that he and his son are both hunters and perhaps one day his grandson will be a hunter, too. Hunting not only offers a source of protein, but also an income, which he has established by turning a constraint into an opportunity.

---

The example of the hunters helps identify community boundaries as semipermeable filters of information with only a few social actors, the hunters among them, able to cross the line and facilitate information flow. The hunters exemplify how social actors are able to use a cultural image to their advantage, taking an environmental condition considered adverse by the mainstream community, and establishing a role in local society within it. By exploring what Allen (1994) would call "possibility space," members of a socially and economically marginal social group have created a means of maintaining their own livelihoods.

### 10.5.5 The Role of Social Actors in Agroenvironmental Change

In addition to the informal conduits of information transmission represented by the hunters, there are also formal agents — the governmental actors — responsible for agricultural extension within the local communities. The interactions between government actors and forest margin communities are historically contingent and locally enacted, yet embedded in government structures and guidelines. The dynamics of the consequent interactions have implications for environmental transformation.

Participation in government agricultural development programs is linked to the recent history of interaction between the government and the communities. The Lampungese community has not participated in government sponsored agricultural development programs, because they see them as encroachments by outsiders into their lifestyle and culture. The Javanese, on the other hand, have participated in various food crop, estate crop, and livestock programs, seeing participation as their duty to follow government rural development initiatives. Nevertheless, being Javanese does not automatically imply that a farmer will accept official agricultural development programs uncritically. Interpersonal relationships among government officials and local farmers, and the relevance of the suggested innovations to the local environment, are crucially important to the success of agricultural development programs, as the following quote from Gauthier (1998) clearly portrays:

> I remember when the present food crops extension worker came to Tegal Mukti. He talked big and told us our ways were no good. He was going to bring us in line with modern agriculture as he had learned it in agricultural school. Before that, we had always waited for the rains to start, around about November, before we planted rice, but he told us we should do it in July. It was a disaster. The harvest was very poor. After that the farmers' groups he'd established broke up, and we don't listen to him anymore. He's completely lost our trust.

There are many other reasons why social actors follow particular development programs, including social standing, power acquisition, and furtherance of their own particular agendas.

In their efforts to build and sustain their livelihoods, take advantage of possibilities, and overcome constraints, farmers engage in complex processes of learning and change, which occur through experimentation, information exchange, mutual learning, and trial and error. Each farmer starts with a set of preferences and attitudes that forms the basis of a strategy. The strategy evolves as information is gained from other actors or acquired by experimentation. A process of mutual learning and adaptation operates within the Lampungese and Javanese communities.

Learning processes involve a variety of different actors. As the distrust of local people for government agricultural representatives has grown through a history of failures and missed opportunities, local, nongovernmental actors have filled this vacuum and become agents of change. Such informal agents of change have had important impact on coping strategies, livelihood trajectories, and environmental transformation within the research area. Some of them gather information; others experiment with and adapt information and practices. Other actors, generally leaders within their respective communities, are catalysts in the dissemination of information (Gauthier, 1998). The interaction of these dynamics results in the emergence of new structural properties within which social actors and the environment interact and adapt.

Through the adaptation and application of new information and practices, agroecosystems are changed incrementally. Over time, transformation of the local environment is driven by rural people's aims for securing a livelihood within an interactive and dynamic framework of social and environmental structures. In transforming the local environment to suit their livelihood needs, farmers create new contexts to which they and others must respond. The effects of the changes are both temporal and spatial.

Coevolutionary environmental change in the area has also impacted upon wildlife populations, providing a final illustration of the dynamics between livelihood strategies, government policy, and environmental change. Environmental transformation in northern Lampung has been marked by a reduction in biodiversity of the natural forest, paralleled by diversification and intensification in agroecosystems, creating new environmental conditions to which the local wildlife had to adjust. In the case of the research area, a change in vertebrate pests has accompanied landscape transformation. In the 1970s, when the landscape was covered by closed canopy rainforest, the most important pests in the area were elephants and tigers. As the forest receded, monkeys and wild pigs became important pests at the forest margins, while paddy rats became the main pests found farther away in the rice fields.

By observing changes in wild animal species' profiles in general, and those of vertebrate pest species in particular, trends in environmental use and related transformation can be traced. The change in pest profile described above indicates environmental changes and predator changes. Where farmers hunt and destroy tigers for their own safety, they remove predator pressure on prey species such as wild pigs. Increased depredation by wild pigs is thus indicative of recent environmental transformation, including changes in predator species activity.

Another example of the link between environmental change and vertebrate pests is the case of the paddy rat (*Rattus argentiventer*). With the spread of wet rice cultivation and the proximity of swamp land, paddy rats have proliferated, causing heavy losses to rice crops. The increase in rat depredation has been paralleled by concerted efforts to reduce snake populations in the paddy fields, and the removal of trees that once served as perches for birds of prey. These changes have created a haven for the paddy rats; their food supply (rice) has increased while predator populations (snakes and birds of prey) have been reduced.

By including wild animals in environmental change research, their part in transformation dynamics becomes clearer. The farmer-wildlife interaction is dynamic and dependent upon local environmental conditions that are continually altered. Because of their mobility, wild animals have faster reaction times to human interventions than do plants and, their consequent impact on human livelihoods are more immediately apparent. Thus local fauna can be used as effective indicators of coevolutionary change.

## 10.6  SOME POLICY IMPLICATIONS OF COEVOLUTIONARY ANALYSES

It is necessary for the policy implications of coevolutionary agroecosystems analysis to be pointed out for two main reasons. In order for interdisciplinary research to be seen as an effective analytical tool by policy makers and program planners, it must be shown to have practical application and relevance. Given the nonlinear coevolutionary agroecological dynamics described in this chapter, the role of conventional policies and programs as blueprints for sustainable rural development is called into question. The assumptions upon which such policies are based should be reviewed if agroenvironmental policies are to become pertinent and effective in the context of changing socioenvironmental conditions.

Coevolutionary agroecological research must be based on intensive field work that describes in detail the lives and daily survival choices of the local people, and their interactions within social communities and biophysical environments (for examples of such work in Latin America, Africa, and Southeast Asia, respectively, see Woodgate, 1992; Campbell, 1998; and Gauthier, 1998). This type of microanalysis provides a basis for situational analysis that is more grounded and better able to ascertain policy impacts (Bryant and Parnwell, 1996). In addition, it can be used by policy makers as a tool for policy formulation and revision, as well as by the social actors who are directly affected by socioenvironmental dynamics. As a tool for policy formulation, microanalysis can inform policy discussion about local specificity, which can be used to adjust and create policy modalities that are locale specific (Trudgill and Richards, 1997) and operate within a broader policy framework, allowing for adjustment at the microlevel. The dialogue between generality and specificity creates a reflexive process which accepts uncertainty as the norm (Wynne, 1992) and can accommodate emerging properties as they arise. Under this scenario the role of local government would be to customize policies to the socioenvironmental specificity of the local area, making socioenvironmental linkages more responsive to local coevolutionary dynamics.

The pertinent points that we will discuss with regard to policies use of an historical perspective in policy formulation and program planning, being aware of differing perceptions of policy aims, and promoting ecological security (Glaeser, 1997) as a basis for policy implementation. Finally, we will argue that the coevolutionary model is a viable and appropriate alternative to neo-liberal development planning.

## 10.6.1 Historical Perspective in Policy Formulation

The coevolutionary approach, based on the microsituations of livelihood decisions and environmental transformations, is highly relevant to agroenvironmental policy formulation and assessment as it traces the links among livelihood, environmental transformation, and policies. It not only looks at the impacts of policies on livelihood and environmental change, but also at how livelihood and environmental change affect policy implementation and outcome. The first ramification of this analysis for policy is the historical context that forms the basis of social actors' acceptance of, and participation in, policy implementation. The historical grounding is pertinent to the social and the environmental contexts. Coevolutionary analysis provides a longitudinal perspective within which interactions between people and the environment are assessed for intervention in environmental systems and the biophysical outcome of such activity as experienced by the individual, the community, and society. History uncovers the social dynamics between officials and policy recipient communities and the key social actors within the community who will influence the outcomes of policy initiatives. Possible sources of conflict between social actors and their historical roots are uncovered. Taken at its fullest, historical analysis could be used for designing policies that redress the social and environmental consequences of past policy failures.

History helps to test policy assumptions with regard to environmental change. For example, Fairhead and Leach (1996) and Scoones (1997) used environmental history to verify assumptions about deforestation and soil degradation in Africa. The field research reported in this chapter elucidates the roles of policy, livelihood, and ecological insecurity in environmental transformation at the forest margin.

## 10.6.2 Differences in Policy Perceptions among Social Actors

Policies represent only one factor in the coevolving dynamics between people and the environment. Policy intervention is understood through experience of past history and through the lens of cultural values and norms. The second aspect of research that is relevant to policy is the way in which it allows understanding and analysis of differences in policy perception among social actors at the field level and within government institutions. Such analysis reveals the diversity of agendas and views within communities and highlights the strategic use of the concept of community when dealing with outside social agents (Leach et al., 1997; Li, 1996) engaged in policy actions.

In the communities studied in North Lampung, certain examples can be derived from empirical evidence, the first being the differing needs of agricultural development policies within the recipient communities. To the Lampungese, current policies favor the Javanese and erode the Lampungese way of life, because they focus on food crop intensification and estate crop plantation monoculture; the Lampungese have traditionally used a tree garden system as their cornerstone. The policy focus on food crops, especially wet rice and sugar cane monocultures, goes against their agricultural lore and enables dissent through nonparticipation and continuation of tree garden cropping systems. These actions of resistance are seen by the Lampungese as ways to differentiate themselves and resist Javanese assimilation.

## 10.6.3 Promotion of Ecological Security

The third issue raised with regard to policies is the role of ecological security (Glaeser, 1997; Langlais, 1995) as a prerequisite to successful implementation of sustainable development policies. Ecological security aims at reducing conflicts by using negotiation and arbitration to resolve differential access to resources. This process has to be carried out through indigenous institutions to maximize the chance of success. To be sustainable, agroenvironmental policies have to include conflict resolution mechanisms to find solutions to problems of access to resources and to reconcile different aims using a variety of approaches (Nepal and Weber, 1995).

The problem of ecological insecurity is illustrated in north Lampung by the land litigation and open conflicts that have arisen in the research area. Mistakes in land boundary registration in the early 1980s turned into open aggression in the mid-1990s. The litigation is complicated by the lack of a clear resolution process and by failure to involve indigenous institutions in the process of negotiation (Gauthier, 1998).

By generating a better understanding of the social dynamics and actors involved, barriers to resolution can be delineated and possible solutions proposed. Resolution of disputes becomes a reflexive, participatory process, which aims at finding solutions, but also, and more importantly, at creating social space in which negotiation and accommodation can evolve. In Lampung, the local government's lack of involvement from the outset and the process which the parties involved in the litigation have to follow reduce the probability of negotiated resolution.

## 10.7 ALTERNATIVES TO NEOLIBERAL POLICIES

Many commentators have called for a middle path to be found for sustainable development that questions the market led, neoliberal approach and avoids utopian visions of an ideal future (Parnwell and Bryant, 1996; Batterbury et al., 1997; Redclift and Woodgate, 1997). The theoretical framework outlined in this chapter is consistent with this call for change, because it can be employed as an analytical tool to critique neoliberal agroenvironmental policies. The application of the holistic approach propounded here is a prerequisite for achieving sustainable agricultural development, and can form the basis of a practical tool to promote sustainable agroecosystems. It can provide a framework for policy and program formulation that takes into account local actors' perspectives and the emergence of new socio-economic and environmental conditions.

Currently, agroenvironmental policies in north Lampung focus on the regulation of individual impacts on the environment through a combination of commodity based policies and environmental restrictions. These policies are implemented through a top-down process, executed by the district government's field staff, and control is maintained through the government's strategy of deconcentration of political power (MacAndrews and Amal, 1993). This keeps decision making power in the hands of the central government. Policies are implemented in a blanket fashion, without incorporating mechanisms for adjusting them to local conditions. The sectoral or commodity focus cannot address the emergent properties that characterize liveli-hood-environment interactions, making prescriptions for sustainable resource use and development less unlikely to prove successful.

Further inhibiting the effectiveness of policies is institutional structure, which is hierarchical and rigid, thus limiting the adaptability which government officials could provide under alternative structural arrangements.

The discord between policy objectives and the needs and desires of the people on the ground, as well as the rigidity of policy implementation, creates opportunity for other social actors to assume functions that are supposed to be carried out by government field workers. The local informal actors are characterized by knowledge of the local situation, flexibility, adaptability to changing conditions, acceptance by the local communities. If these characteristics were considered by government rep-resentatives, the government would be better able to contribute to the sustainability of rural development interventions.

In contrast to the government's approach in Lampung, the coevolutionary approach highlights the agendas of participating social actors. Officials following this approach become facilitators in conflict resolution negotiations and the devel-opment of sustainable strategies for environmental use. Uncertainty is inherent in the socioenvironmental system, and heterogeneity and flux are the norms in complex socioenvironmental settings. The coevolutionary approach shows that the top-down, centrally planned approach has little chance of success because the interactions among policies, social structures, actors, livelihoods, and the environment are cha-otic. In addition, the biophysical setting for policy implementation is altered con-tinuously by the other factors. That leads to consequences unanticipated by policy makers and program planners.

To be effective, agroenvironmental policies should have shorter feedback loops (Woodgate, 1992; Gauthier, 1998) in order for implementation lessons to be learned and policy adjustments implemented; the policy development process must be acces-sible, equitable, and adaptive (Sahl and Bernstein, 1995). The type of policy man-agement we envision is reflexive and views social actors as involved agents for change in social and ecological structures; they are stakeholders in the environment and the policies that impact it, follow ecological and policy guidelines, and produce environ-mental and policy outcomes. This duality is not recognized by policies based on the assumptions that implementation is a one way process and that a single cause and effect chain is possible. Neoliberal policy is more dependent on the balance of nature view, whereas coevolutionary policy would be based on what new paradigm ecologists (Pahl-Wostl, 1995; Pickett et al., 1997) would call the flux of nature. Without this two way flow, necessary lessons go unlearned and sustainability is compromised.

The review of policies from a coevolutionary perspective supports the call by some policy analysts to base policies that affect the environment and people's interactions on subsidiarity and the precautionary principle (Trudgill, 1990; Trudgill, 1992) and to shift from normal science and management to post normal science (Funtowicz and Ravetz, 1993). It also suggests a constant questioning, monitoring, and modification of policy and its implementation.

This call for a post-normal paradigm should be extended to policy makers and technocrats to facilitate and complement local people's knowledge and practices.

Only in this way is it possible to achieve locally sustainable environmental uses. Under the post-normal paradigm, policies become tools to facilitate problem resolution at the local level. It encourages political and scientific debate that recognizes the needs of local and national stakeholders, thus making policy formulation, implementation, and evaluation more reflexive, smaller scale, and adaptive to constantly changing socioenvironmental dynamics.

## REFERENCES

Adam, B., Time and the environment, in Redclift, M.R. and Woodgate, G., Eds., *The International Handbook of Environmental Sociology*, Edward Elgar, Cheltenham, UK, 1997.

Allen, P.M., Evolution, sustainability and industrial metabolism, in Simonis, R.U. and Simonis, U.S., Eds., *Industrial Metabolism: Restructuring for Sustainable Development*, United Nations University Press, Tokyo, 1994.

Arnold, J.E.M. and Dewees, P.A., *Farms, Trees and Farmers: Responses to Agricultural Intensification*, Earthscan Publications, London, 1997.

Batterbury, S., Forsyth, T., and Thompson, K., Environmental transformations in developing countries: Hybrid research and democratic policy, *Geographical J.*, 163(2). 126–132, 1997.

Biro Pusat Statistik, Lampung dalam Angka, *Kantor Statistik*, Propinsi Lampung, Bandar Lampung, Indonesia, 1994–1995.

Bridges, E.M., *World Soils*, 2nd ed., Cambridge University Press, Cambridge, UK, 1978.

Bryant, R.L. and Parnwell, M.J.G. Introduction: Politics, sustainable development and environmental change in South-East Asia, in Parnwell, M.J.G. and Bryant, R.L., Eds., *Environmental Change in South-East Asia: People, Politics and Sustainable Development*, Routledge, London, 1996.

Campbell, M.O., Interactions between Biography and Rural Livelihoods in the Coastal Savannah of Ghana, Ph.D. thesis, Wye College, University of London, London, 1998.

Clark, J.J., *Nature in Question: an Anthology of Ideas and Arguments*, Earthscan Publications, London, 1993.

Conway, G.R., *After the Green Revolution: Sustainable Agriculture for Development*, London, Earthscan, 1990.

Dickens, P., Beyond sociology: Marxism and the environment, in Redclift, M.R. and Woodgate, G., Eds., *The International Handbook of Environmental Sociology*, Edward Elgar, Cheltenham, UK, 1997.

Engels, F., *The Dialectics of Nature*, Progress Publisher, Moscow, 1959, 12.

Fairhead, J. and Leach, M., *Misreading the African Landscape: Society and Ecology in a Forest-Savannah Mosaic*, Cambridge University Press, Cambridge, UK, 1996.

Fiedler, P.L, White, P.S., and Leidy, R.A., The paradigm shift in ecology and its implications for conservation, in Pickett, S.T.A., Ostfeld, R.S., Shachak, M., and Likens, G.E., Eds., *The Ecological Basis of Conservation: Heterogeneity, Ecosystems and Biodiversity*, Chapman and Hall, New York, 1997.

Funtowicz, S.O. and Ravetz, J.R., Science for the Post-Normal Age, *Futures*, 25, 739–755, 1993.

Gandy, M., Postmodernism and environmentalism: complementary or contradictory discourses? in Redclift, M.R. and Woodgate, G., Eds., *The International Handbook of Environmental Sociology*, Edward Elgar, Cheltenham, UK, 1997.

Gass, G., Biggs, S., and Kelly, A., Stakeholders, science, and decision making for poverty-focused rural mechanization research and development, *World Dev.*, 25(1), 115–126, 1997.

Gauthier, R.C.T., Policy, Livelihood and Environmental Change at the Forest Margin in North Lampung, Indonesia: A Coevolutionary Analysis, Ph.D. thesis, Wye College, University of London, London, 1998.

Giddens, A., *Central Problems in Social Theory: Action, Structure and Contradiction in Social Analysis*, Macmillan, London, 1979, 69.

Giddens, A., *The Constitution of Society: An Outline of the Theory of Structuration*, Polity Press, Cambridge, UK, 1984.

Glaeser, B., Environment and developing countries, in Redclift, M.R. and Woodgate, G., Eds., *The International Handbook of Environmental Sociology*, Edward Elgar, Cheltenham, UK, 1997.

Gliessman, S.R., Ed., *Agroecology: Researching the Ecological Basis for Sustainable Agriculture*, Springer-Verlag, London, 1990, 366–367.

Hadikusuma, H.H., *Masyarakat dan Adat-Budaya Lampung*, Mandar Maju, Bandung, Indonesia, 1989.

Langlais, R., Reformulating security: A case study from arctic Canada, *Humanekologiska Skrifter*, Götenborg University, Götenborg, 13, 1995.

Leach, M., Mearns, R., and Scoones, I., Institutions, consensus and conflict: implications for policy and practice, *IDS Bulletin*, 28(4), 90–95, 1997.

Levang, P., Farming systems and household incomes, in Pain, M., Ed., *Transmigration and Spontaneous Migration, Propinsi Lampung, Sumatra, Indonesia*, ORSTOM, Bondy, France, 1989.

Li, T.M., Images of community: discourse and strategy in property relations, *Dev. Change*, 27(3), 501–527, 1996.

MacAndrews, C. and Amal, I., Hubungan Pusat-Daerah dalam Pembangunan, Raja Grafindo Persada, Jakarta, 1993.

McIntosh, R.P., Pluralism in ecology, *Ann. Rev. Ecol. Syst.*, 18, 321–341, 1987.

Marx, K., *Early Writings*, Colletti, M., Ed., Penguin, Harmondsworth, UK, 1975, 355.

Meyer, W., *Human Impact on the Earth*, Cambridge University Press, Cambridge, 1996.

Nepal, S.K. and Weber. K.E., Managing resources and resolving conflicts: national parks and local people, *Int. J. Sustain. Dev. World Ecol.*, 2, 11–25, 1995.

Norgaard, R.B., Coevolutionary agricultural development, *Economic Dev. Cultural Change*, 32(3), 525–546, 1984.

Norgaard, R.B., *Development Betrayed: The End of Progress and a Coevolutionary Revisioning of the Future*, Routledge, London, 1994, 530.

Norgaard, R.B., A coevolutionary environmental sociology, in Redclift, M.R. and Woodgate, G., Eds., *The International Handbook of Environmental Sociology*, Edward Elgar, Cheltenham, UK, 1997.

Pahl-Wostl, C., *The Dynamic Nature of Ecosystems: Chaos and Order Entwined*, John Wiley & Sons, Chichester, UK, 1995.

Pain, M., Benoit, D., Levang, P., and Sevin, O., *Transmigration and Spontaneous Migration in Indonesia: Propinsi Lampung*, ORSTOM, Bondy, France, 1989.

Parnwell, M.J.G. and Bryant, R.L., Conclusion: Towards sustainable development in South-East Asia? in Parnwell, M.J.G. and Bryant, R.L., Eds., *Environmental Change in South-East Asia: People, Politics and Sustainable Development*, Routledge, London, 1996.

Pearce, D., Markandya, A., and Barbier, E., *Blueprint for a Green Economy*, Earthscan Publications, London, 1989.

Pemerintah Propinsi Daerah Tingkat I-Lampung, *Pola Pelaksanaan Transmigrasi Umum, Resettlement Transmigrasi Lokal di Daerah Lampung*, Pemerintah Propinsi Daerah Tingkat I, Lampung, Bandar Lampung, Indonesia, 1983.

Pemerintah Propinsi Daerah Tingkat I-Lampung, *Rencana Pembangunan Lima Tahun Keenam*, Pemerintah Propinsi Daerah Tingkat I, Lampung, Bandar Lampung, Indonesia, 1995.

Pickett, S.T.A., Ostfeld, R.S., Shachak, M., and Likens, G.E., *The Ecological Basis of Conservation: Heterogeneity, Ecosystems and Biodiversity*, Chapman and Hall, New York, 1997.

Redclift, M.R. and Woodgate, G., Sustainability and social construction, in Redclift, M.R. and Woodgate, G., Eds., *The International Handbook of Environmental Sociology*, Edward Elgar, Cheltenham, UK, 1997.

Sage, C.L., The search for sustainable livelihoods in transmigrant settlements, Indonesia, in Parnwell, M. and Bryant, R., Eds., *Environmental Change in South-East Asia: People, Politics and Sustainable Development*, Routledge, London, 1996.

Sahl, J.D. and Bernstein, B.B., Developing policy in an uncertain world, *Int. J. Sustain. Dev. World Ecol.*, 2(2), 124–135, 1995.

Scoones, I., The dynamics of soil fertility change: historical perspectives on environmental transformation from Zimbabwe, *Geographical J.*, 163(2), 161–169, 1997.

Sprugel, D.G., Disturbance, equilibrium and environmental variability: what is "natural" vegetation in a changing environment? *Biological Conserv.*, 58, 1–18, 1991.

Thomas, D and Middleton, N., *Desertification: Exploding the Myth*. John Wiley & Sons, Chichester, UK, 1994.

Trudgill, S.T., *Barriers to a Better Environment*, Belhaven, London, 1990.

Trudgill, S.T., Environmental issues, *Processes in Physical Geography*, 16, 223–229, 1992.

Trudgill, S.T. and Richards, K., Environmental science and policy: generalizations and context sensitivity, *Trans. Inst. Br. Geogr.*, 22(1), 5–12, 1997.

Whitten, A.J., Damanik, S.J., Anwar, J., and Hisyam, N., *The Ecology of Sumatra*, Gadjah Mada University Press, Yogyakarta, Indonesia 1987.

Woodgate, G., Sustainability and the Fate of the Peasantry: The Political Ecology of Livelihood Systems in an Upland Agroecosystem in the Central Highlands of Mexico, Ph.D. thesis, Wye College, University of London, London, 1992.

Wynne, B., Uncertainty and environmental learning: reconceiving science and policy in the preventive paradigm, *Global Environmental Change*, 2, 111–127, 1992.

Yusuf, T., *Profil Propinsi Lampung*, Gunung Pegasi, Bandar Lampung, Indonesia, 1992.

CHAPTER **11**

# Operationalizing the Concept of Sustainability in Agriculture: Characterizing Agroecosystems on a Multi-Criteria, Multiple Scale Performance Space

Mario Giampietro and Gianni Pastore

## CONTENTS

## 11.1 INTRODUCTION

Agriculture operates on the interface of two complex, hierarchically organized systems: socioeconomic systems and natural ecosystems (Hart, 1984; Conway, 1987; Lowrance et al., 1986; Ikerd, 1993; Giampietro, 1994a,b, 1997a; Wolf and Allen, 1995). This implies that in any analysis of a defined farming system one will always find legitimate and contrasting perspectives with regard to the effects of changes in the system (Giampietro, 1999). For example, increasing return for farmers (intensification of crop production) can be coupled with more stress on ecological systems (loss of biodiversity and soil erosion). Similarly, improvements for certain social groups (lower retail price of food for consumers) can represent a step back for others (lower revenues for farmers).

The implications are that changes in agriculture, induced by new policies, technical innovations, or sudden changes in ecological boundary conditions, are unlikely to result in improvements or worsen when considering the various perceptions of various stakeholders (defined as social actors affected by and affecting events). For example, the introduction of mechanical power in agriculture (which represented a tremendous boost in the ability of humans to transport goods and people, till soil, and pump water for irrigation) implied the disappearance of jobs and revenues related to animal powered activities. The generation of winners (in certain social groups) was coupled to the generation of losers. In the same way, nonequivalent descriptions of changes in agriculture referring to different space-time scales (soil, farm fields, watersheds, regions, the world) can imply the detection of different (side) effects induced by the process of agricultural production. For example, large scale conversion of the natural landscape into crop production systems based on monoculture is likely to induce a negative effect on biodiversity and/or stability of water cycles on a large scale. These effects cannot be easily "guessed" when evaluating the influence of monoculture on a single crop field.

When dealing with the issue of sustainability, a correct assessment of agricultural performance should be based on an integrated analysis of trade-offs rather than on the use of reductionistic analyses searching for optimal solutions (Optimal for whom? Optimal for how long? Optimal on which scale?). An analysis of agricultural performance should be based on an integrated set of indicators that are able to

(1) reflect various perspectives and (2) read the changes occurring on different hierarchical levels in parallel on space-time scales. This is the only way to usefully characterize the effects that a proposed technological or policy change can be expected to induce in the various actors involved and in relation to processes occurring on different scales.

The theoretical discussion in this chapter will be complemented by practical examples taken from a case study. We will use the findings of a four year project aimed at characterizing the effects on sustainability of the process of intensification of production in rural areas of China. The complete results of this study are presented in four papers (Giampietro et al. 1999; Li Ji et al., 1999; Giampietro and Pastore, 1999; Pastore et al., 1999) to which we refer the reader for more detailed explanations of data and methods.

## 11.2 THEORETICAL BASIS OF THE INTEGRATED ASSESSMENT APPROACH

### 11.2.1 Nested Hierarchical Systems and Nonequivalent Descriptive Domains

Agricultural systems are complex systems made up of many different components that operate in parallel on different space-time scales. These components include soil microorganisms, populations of selected plant species in crop fields, individual farmers, farmer households, rural communities, local economies, local agroecosystems, watersheds, regional economies, biospheric processes stabilizing, bio-geochemical cycles of water and nutrients, and socioeconomic processes operating at the macroeconomic level stabilizing the boundary conditions of farming activities. In addition to being hierarchically organized on several scales, ecological and human systems are made up of "holons" (Koestler, 1968; 1969). A holon is a whole consisting of smaller parts (as a human being is made of organs, tissues, cells, molecules, etc.) which forms a part of some greater whole (as an individual human being is part of a household, a community, a country, the global economy).

All natural systems of interest for sustainability (i.e., biological systems and human systems analyzed at different levels of organization and scales above the molecular one) are "dissipative systems" (Glansdorf and Prigogine, 1971; Nicolis and Prigogine, 1977; Prigogine and Stengers, 1981). They are self organizing, open systems, operating away from thermodynamic equilibrium. In order to remain alive or integrated they have to be able to stabilize their own metabolism within their given context. Put in another way, living systems have to make available an adequate amount of food, and economic systems have to make available an adequate amount of added value, as well as an adequate amount of material and energy input. Because of this forced interaction with their context, dissipative systems are necessarily open and therefore "becoming" systems (Prigogine, 1978). This implies that they (1) are operating in parallel on several hierarchical levels (various patterns of self organization can be detected only by adopting different space-time windows of observation) and (2) are changing their identity in time at different rates over their various levels of organization. The concept of self organization in dissipative systems is deeply linked to the ideas of parallel levels of organization on different space-time scales and evolution.

Various authors have defined hierarchical systems in a way that is consistent with the foregoing discussion. According to O'Neill (1989), a dissipative system is hierarchical when it operates on multiple spatiotemporal scales when different process rates are found in the system. Simon writes that, "Systems are hierarchical when they are analyzable into successive sets of subsystems" (1962). Another definition is proposed by Whyte: "A system is hierarchical when alternative methods of description exist for the same system" (1969). These definitions point to this conclusion: the existence of different levels and scales at which a hierarchical system can be analyzed implies the existence of nonequivalent descriptions of it.

For example, we can describe a human being at the microscopic level to study the cellular processes occurring within his body. When we look at a human at the cellular scale we can take a picture of him with a microscope (Figure 11.1a). This type of description is not compatible with the description of the same human being's face, e.g., the description needed when applying for a driving license (Figure 11.1b). No matter how many pictures we take with a microscope of a defined human being, the type of pattern recognition of that person at the cellular level will not be

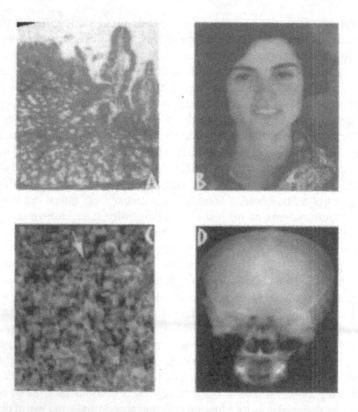

**Figure 11.1** Nonequivalent descriptive domains needed to obtain nonequivalent pattern recognition in nested hierarchical systems.

equivalent to the description of the human being at the organismal level (Figure 11.1b). The ability to detect the identity of the face of a given person is an emergent property linked to a description which is in turn linked to a defined space-time window. The face cannot be detected using a description linked to a very small space-time window (the scale used for looking at individual cells), just as it cannot be detected using a description linked to a larger scale (a scale used for looking at social relations, exemplified by Figure 11.1c).

It should be noted that the term "emergent property" can be misleading. The term does not refer to the analyzed system itself, but rather to the need for a pattern recognition in relation to an assigned goal for the description. When dealing with a system organized hierarchically, it does not make sense to speak of pattern recognition. There are an infinite number of patterns overlapping across scales waiting for recognition within every self-organizing adaptive hierarchical system. We take a photograph able to detect a face when we need input for a driving license, and we make an X-ray image of the same head when we are looking for an input in a medical investigation (Figure 11.1d). The four recognizable patterns shown in Figure 11.1 are present in parallel at any time. We simply choose to look at the system in a particular way, and this choice leads us to focus on one pattern (or scale, or space-time window) rather than the others (Giampietro, 1999).

Human societies and ecosystems are generated by processes operating on several hierarchical levels over a cascade of different scales. They are perfect examples of dissipative hierarchical systems that require many nonequivalent descriptions, used in parallel, to analyze their relevant features in relation to sustainability (Giampietro 1994a; 1994b; 1997c; 1999; Giampietro et al., 1997; Giampietro et al., 1998a; 1998b; Giampietro and Pastore, 1999). Using the epistemological rationale proposed by Kampis for defining a system as "the domain of reality delimited by interactions of interest" (1991), we can introduce the concept of descriptive domain in relation to the analysis of a system organized on nested hierarchical levels. A descriptive domain is the domain of reality resulting from an arbitrary decision to describe a system in relation to (1) a defined set of encoding variables to catch a selected set of relevant qualities linked to the choice of variables and (2) a defined space-time horizon for the behavior of interests determined by the resulting relevant space-time differential (needed to detect and characterize the behavior of interest in terms of a dynamic generated by an inferential system over a set of variables linked to a pattern recognition obtained when referring to a particular hierarchical level). The very definition of a boundary for the system (linked to the previous selection of a given time horizon) will affect the identity of the differential equations used to simulate the behavior of interest in relation to a particular selection of variables (Rosen, 1985).

To clarify this concept we can reconsider the four views of the same system shown in Figure 11.1, using a metaphor of sustainability. Imagine that the four nonequivalent descriptions presented in Figure 11.1 portray a country (e.g., The Netherlands) rather than a person. We can easily see how the parallel use of different descriptive domains is required to obtain an integrated analysis of the country's sustainability. For example, looking at socioeconomic indicators of development we see satisfying levels of GNP and good indicators of equity and social progress, just as we see an attractive woman in Figure 11.1b. These qualities of the system are required to keep the stress on social processes low. If we look at the same system and use different encoding variables (e.g., biophysical variables) we can see a few problems not detected by the previous selection of encoding variables; such as accumulation of excess nitrogen in the water table, growing pollution in the environment, and excessive dependency on fossil energy and imported resources for the agricultural sector — just as the description in Figure 11.1d may allow us to see sinusitis and dental problems. This comparison demonstrates that even when the same physical boundary and scale for the system are maintained, a different selection of encoding variables can generate a different assessment of the performance of the system.

The process becomes more difficult when we decide to use other indicators of performance that must relate to descriptive domains based on different space-time differentials. For example, we could analyze the sustainability of Dutch agriculture using a scale equivalent to Figure 11.1a. In this analysis, related to lower level components of the system (which require for their description a smaller space-time differential), we might be concerned with measuring technical coefficients (e.g., input/output) of individual economic activities. Clearly, this knowledge is crucial for determining the viability and sustainability of the whole system because it relates to the possibility of improving or adjusting the overall performance of Dutch economic processes if and when changes are required. In the same way, an analysis of the relations of the system with its larger context implies the need for a descriptive domain based on larger scale pattern recognition, equivalent to Figure 11.1c. For The Netherlands, this could be an analysis of institutional settings, historical entailments, or cultural constraints over possible evolutionary trajectories.

In conclusion, when dealing with the sustainability of complex adaptive systems, the existence of irreducible relevant behaviors expressed in parallel over various relevant space-time differentials implies a need for using different descriptive domains in parallel. This claim has two important implications:

1. It is impossible for practical reasons to handle the amount of information that would be required to describe the sustainability problems. Any specific description, based on the handling of a finite information space, misses relevant information about the system.
2. It is impossible for theoretical considerations to collapse the complexity of an adaptive system organized over several relevant hierarchical levels into a simple model based on a single formal inferential system (Rosen, 1985; 1991). After accepting that qualities detectable only within different descriptive domains can be reflected only by using nonequivalent models, we are forced to accept that these models are not reducible to each other.

## 11.2.2 Examples from Agricultural Analyses

Understanding the holarchic structure of agricultural systems is a fundamental prerequisite for a sound analysis of their performance. Policy suggestions based on agricultural research tend to be plagued by systematic errors in the structuring of the problem through models. In practice, scientific analyses are based on only one hierarchical level of analysis, and as a consequence, have to use encoding variables belonging to only one descriptive domain. As a result of this method, analyses performed at a certain level in relation to a certain issue (e.g., compatibility of crop production techniques with soil health) do not necessarily provide sound information on what goes on at other levels in relation to distinct issues (e.g., compatibility of the production technique with expected farmer income in a defined rural community operating in a given socioeconomic context) (Giampietro, 1994a, 1997a, 1997b, 1999).

The choice of a multicriteria, multilevel representation of performance over distinct descriptive domains is a required choice when dealing with sustainability. Without using a multilevel analysis, it is very easy to devise models that simply suggest shifting a particular problem between different descriptive domains. Optimizing models, which are based on a simplification of real systems within a single descriptive domain, tend to externalize the analyzed problem out of their own boundaries (e.g., economic profit can be boosted by increasing ecological or social stress; ecological impact can be reduced by reducing economic profit, and so on). When the use of such models predominates, policy suggestions are based on the detection by a model of some "benefits" on certain descriptive domains and the ignoring of some "costs" detectable only on different descriptive domains. This problem, faced by all monocriterial analyses, can be avoided by the parallel use of nonequivalent indicators belonging to different relevant and complementing descriptive domains, which makes it possible to easily detect such "epistemological cheating." Problems externalized by one model on a given scale (e.g., describing items in economic terms over a 10-year time horizon) will reappear amplified in one of the parallel models (e.g., when describing the same change in biophysical terms or on a larger time horizon).

As noted in the example shown by Figure 11.1, the ability of any model to see and encode some qualities of the natural world implies that the same model cannot see other qualities detectable only on different descriptive domains. A simple practical example dealing with historical changes in a farming system serves to clarify this point.

Farming systems in rural China have undergone dramatic changes in recent decades. Figure 11.2 shows four nonequivalent indicators that can be used to characterize these changes.

### 11.2.2.1 The Farmers' Perspective

The first indicator in Figure 11.2a is related to the profile of land use. This assessment indicates the percentage of crop land used to guarantee an adequate supply of nitrogen for crop production. In the 1940s, about 30% of crop land was allocated to green manure cultivation and was unavailable for subsistence or cash crop production. The intensification of crop production, driven by population growth and socioeconomic pressure, led to a progressive abandonment of the use of green manure (too expensive in terms of land and labor demand) in favor of synthetic fertilizer. This shift resulted in a sensible increase in multiple cropping practices and a dramatic improvement in agronomic indices of crop yield per

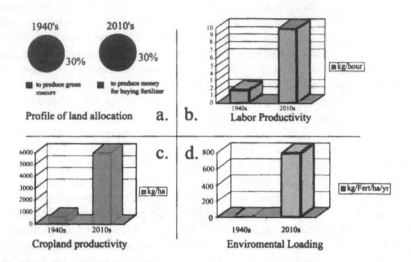

**Figure 11.2** Different indicators that can be used to characterize historical trends in rice farming in China.

hectare. This dramatic increase in crop production led toward self sufficiency and freed land for cultivation of cash crops (Li Ji et al., 1999). Current trends show an increase in demographic and economic pressures leading to further intensification of agricultural throughputs (Giampietro, 1997a; 1997b), which will likely, by 2010, bring the percentage of land allocated to producing adequate nitrogen back to the 30% mark, where it was in the 1940s. About 30% of the land invested in cash crops will be used just to pay for fertilizer inputs. When considering how much land is required for stabilizing agricultural production, both solutions require a 30% investment of the total budget of available land and are thus equal for the farmer. According to the farmers' view, the same fraction of land is lost, whether it is to green manure production or to crop production to purchase chemical fertilizer. The characterization (mapping of system qualities) given in Figure 11.2a does not distinguish the differences implied by these two solutions. Other criteria and other indicators are needed if we want to obtain a better explanation of such a trend.

### 11.2.2.2 The Households' Perspective

When considering as an indicator of performance the productivity of labor (Figure 11.2b) we see that the chemical fertilizer solution implies a much higher labor productivity than the green manure solution. Higher labor productivity translates into a higher economic return for each unit of labor. Depending on the budget of working time available to the household, it is possible to reduce the fraction of working time allocated to self-sufficiency and increase the fraction of working time allocated to cash flow generation and leisure. Farmers will prefer the chemical fertilizer solution because it allows a better allocation of their time.

## 11.2.2.3 The Nation's Perspective

When considering cropland productivity as performance indicator (Figure 11.2c), we see that the chemical fertilizer solution implies much higher land productivity than the green manure solution. The land used to produce crops for the market to pay for chemical fertilizer is perceived as lost by farmers. At the national level, it is seen as land that produces food for the urban populations. Green manure production is seen as use of crop land without generating food. The goal of the government of China to boost the food surplus in rural areas to feed the growing urban population may actually lead to policies of intensification of agricultural production through further increases in technical inputs. This goal might increase the fractions of farmers' lands budgets needed to meet the cost of purchasing additional chemical fertilizers, a result that would discourage farmers from inten-sifying their use of technical inputs. If this became the case, the central government could decide to subsidize the use of these inputs, lowering the cost of fertilizer and reducing the fraction of land that farmers have to use for procuring fertilizer. That would change the situation from the farmers' perspective, and induce an intensification of agricultural production. The reduction of land lost to buy chem-ical fertilizer (as detected by the farmers' perception) and the increase in cropland productivity (as detected through the central government's perception), both obtained by subsidization of fertilizer, adds another variable — the economic cost of internal food production. The advantage provided by the use of fertilizer subsidies — characterized as "cropland productivity" — induces a side effect which can be detected only by using an additional criterion at the national level: the economic burden of subsidizing technical inputs. Note that this indicator is not shown in Figure 11.2.

## 11.2.2.4 The Ecological Perspective

From the ecological perspective, we find different consequences of the two solutions allocating 30% of land to nitrogen maintenance. The use of green manure in the 1940s was benign to the environment because the flow of nutrients in the cropping system was kept within a range of values of intensity close to those typical of natural flows. In contrast, the acceleration of nutrient throughputs induced by the use of synthetic fertilizers dramatically increased the environmental stress on the agroec-osystems. Therefore, when biophysical indicators of environmental stress are used to characterize the changes in rural agriculture in China (Figure 11.2d), we obtain an assessment of performance that is unrelated to and logically independent from assessments based on the use of economic variables; it shows that the synthetic fertilizer solution is not conducive to healthy soil.

## 11.2.2.5 Lessons from This Example

This example demonstrates several points. The same criteria (land demand per output) can require different indicators to reflect different hierarchical levels. The indicators in Figure 11.2a and Figure 11.2c show contrasting indications of the green manure solution and the synthetic fertilizer solution in relation to use of land. From the farmers' perspective, there is no difference in the two solutions, but they are dramatically different from the national perspective.

Criteria and indicators referring to different descriptive domains (such as environmental loading assessed in kg of fertilizer/ha versus labor productivity expressed in kg of crop/hour) reflect not only incommensurable qualities, but also unrelated systems of control. As a consequence, when dealing with trade-offs defined on different descriptive domains, we cannot expect to establish simple protocols of optimization to compare and maximize relative costs and benefits.

## 11.3 INCOMMENSURABLE SUSTAINABILITY TRADE-OFFS

### 11.3.1 Multi-Criteria Analysis and Incommensurable Indicators of Performance

Multi-criteria methods of evaluation are gaining attention among the economic community (Bana e Costa, 1990; Nijkamp et al., 1990; van den Bergh and Nijkamp, 1991; Munda et al., 1994). Multi-criteria evaluation has demonstrated its usefulness in conflict management for many environmental management problems (Munda et al., 1994). The major strength of multi-criteria methods is their ability to address problems marked by various conflicting evaluations. In general, a multi-criteria model presents the following two aspects:

1. There is no solution optimizing all the criteria at the same time, and therefore decision making implies finding compromise solutions.
2. The relations of preference and indifference are inadequate; when one action is better than another according to some criteria, it is usually worse according to others. Many pairs of actions remain incompatible with respect to a dominant relation.

The basic idea of a multi-criteria analysis is linked to a characterization of system performance based on a set of aspects/qualities, none of which can be expressed as functions of the others. They are nonequivalent and nonreducible. When such a characterization is realized in a graphic form, it is possible to have an overall assessment of system performance through a visual recognition of the difference between the profile of expected or acceptable values and the profile of actual values over families of indicators of performance. The various families of indicators should be able to catch noncomparable qualities expressed by variables belonging to nonequivalent descriptive domains.

This method of analysis is quite old; it is used, for example, in marketing (e.g., spider web analysis) for assessment of consumer satisfaction. Wide differences between expected and actual values indicate lack of consumer satisfaction, and areas of the graph in which the gap between expectation and actual performance is wide indicate priorities in terms of intervention. Such a graphic analysis is illustrated in Figure 11.3. The subject of this figure — consumer satisfaction with a new model of automobile — is related to the issues of agricultural sustainability. The new car

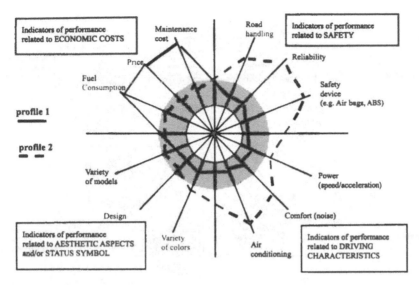

**Figure 11.3** Example of integrated assessment based on incommensurable criteria: Consumer satisfaction with two models of cars.

model will not be sustainable in the market place if it fails one of the qualities affecting consumer choice, no matter how well it performs on the other parameters (Giampietro, 1999).

In the field of natural resource management, the same approach has been proposed under the acronym AMOEBA by Brink et al. (1991) as a tool for dealing with the multidimensionality of environmental stress assessment. Brink et al. propose the use of different indicators of ecological stress belonging to descriptive domains linked to different space-time scales.

## 11.3.2 The Multi-Criteria, Multiple Scale Performance Space

In our approach, the graphic representation of the system is based on a division of a radar diagram into four quadrants, each describing a distinct perspective (Figure 11.4). Within each quadrant, a number of axes representing different indicators of performance are drawn. The choice of quadrants and axes is arbitrary and based on characteristics of the system considered relevant for the analysis. This value call opens the door to participatory techniques that should be adopted when using this method of analysis. Returning to the example of the car in Figure 11.3; no one can decide what is the optimal design for a car without asking potential drivers about their specific expectations and needs. This simple analogy suggests that a group of experts cannot decide from their desks what is the optimal system of production for a defined crop or farming system without checking the compatibility of their assumptions with the farmers who are expected to adopt the system.

When building a multi-criteria, multiple scale performance space (MCMSPS) with regard to agricultural sustainability, the main aspects to be considered are those

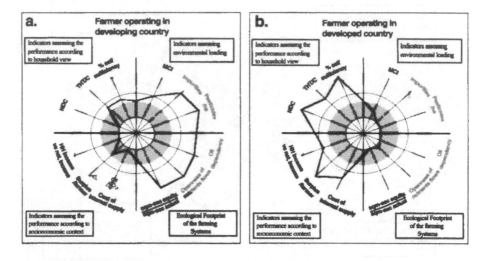

**Figure 11.4** Examples of multi-criteria, multiple scale performance spaces.

characterizing the activity of farming in relation to its socioeconomic context (economic viability and social acceptability) and ecological context (ecological compatibility and congruence between the requirements for and the availability of natural resources) (Giampietro et al., 1994; Giampietro, 1997a; 1997b; Giampietro and Pastore, 1999; Giampietro, 1999).

In the examples provided in Figure 11.4, the agricultural system is described by quadrants that refer to the following aspects of performance: benefits and costs to the farmer or household (upper left quadrant), role in the national or regional economy (lower left quadrant), the extent of local environmental loading (upper right quadrant), and the requirements for natural resources compared to the availability of the resources (lower right quadrant). The latter is a measure of the extent to which a steady state description of the agricultural system (the one used when drawing a boundary around the system of production) misses relevant information.

The lower right quadrant accounts for the fact that today almost no agricultural system is either closed or in a steady state. The inputs and outputs involved in describing matter and energy flows in production systems are increasingly based on stock depletion (of fossil fuels, underground water, soil, and biodiversity) and filling of sinks (accumulation of pesticides in the environment and nitrogen in the water table, etc.). The physical boundaries used to define a farm no longer coincide with the ecological footprint of the process of production inputs (such as feed used in animal production); the inputs are often imported from elsewhere to boost the productive capacity of farmers. The flows of added values, matter, and energy required to generate the inputs do not necessarily coincide in space.

Figure 11.4 represents the effects of changes in the system in parallel on different hierarchical levels (descriptive domains related to different space-time scales) and according to any given perspective selected among a virtually infinite number of possible indicators.

## 11.4  THE CHALLENGES IMPLIED BY A COMPLEX REPRESENTATION OF REALITY

### 11.4.1  Acknowledging the Evolutionary Nature of Agriculture

Numerical assessments obtained after selecting a set of indicators (such as the ones reported in Figure 11.2) should be seen as snap shot pictures of the farming system under analysis. They can be used to explain possible combinations of land and labor allocation profiles, reflecting a given set of boundary conditions, such as yields, prices, area of crop land, and existing government regulations. Therefore, any analysis based on these assessments has to follow the *ceteris paribus* assumption: the system has to be in a quasi-steady state to be characterized with numerical indicators.

Agricultural systems evolve in time over all their different scales, as illustrated in Figure 11.5. The parallel functioning on several scales of the system implies that the values of a particular set of variables (e.g., a household) are forced into congruence with the values of other sets of variables read on different hierarchical levels (e.g., the economic context within which the household is operating). For example, the economic return of farm labor (in local currency per hour) as seen at the farming-system level affects the cost of food for the urban population (in percent of income spent on food) as seen on the national level. In the same way, land productivity in terms of kg of output per hectare as seen at the farming system level affects the value of environmental loading at the soil level (kg of nitrogen fertilizer applied per hectare per year) or village level (concentration of nitrates and phosphates in the water table).

Each of the various holons that can be distinguished in the system (e.g., households, villages, the nation) has a different set of goals expressed in a particular sets

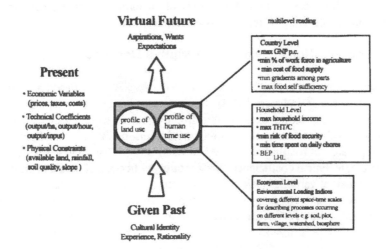

**Figure 11.5** Evolutionary trajectory between a given past and a virtual future through viable states.

of variables. Within their level specific description of farming, the actors will look for the best combination of profiles of human time and land allocation (the combination that satisfies the performance of farming according to their own perspectives). The goals and boundary conditions of different levels of the system do not coincide. This means that holons belonging to different levels have distinct views of what represents a satisfying allocation of their resources.

It is here that the holarchic structure of the system enters into play. Different holons, even those belonging to different hierarchical levels and having different views with good and bad results still belong to the same system and affect each other. This implies that the choice of each holon will affect choices of other level holons and vice versa. What is considered bad in the short term (e.g., paying taxes for farmers) can become good in the long-term by inducing positive effects on processes occurring at different scales (e.g., making it possible for the country to provide better health services for rural communities). The national government (a higher level holon) can change price policies in agriculture or establish new laws and regulations to influence choices of farmers (lower level holons). In response, farmers can change their behavior, for example, by reducing the share of their work time allocated to farming in favor of off farm labor. Agroecosystems can react through reduced crop output due to loss of soil fertility and salinization.

The complex nature of agroecosystems implies that a certain tension always exists among the different levels of the system. Within the same holarchy, contrasting perspectives of different holons are not only inevitable but are necessary for long-term stability (Giampietro 1994a). Given this built in tension, holons belonging to different levels must be capable of continuously negotiating compromise solutions. The various holon specific satisfying solutions and the various perspective dependent assessments of agricultural performance must continuously confront each other.

To enable this process of continuous definition and negotiation of satisfying courses of action, scientific representations of sustainability issues have to reflect such a complex structure of relations (Giampietro, 1999).

## 11.4.2 Bridging Nonequivalent Descriptive Domains

The graphic representation in Figure 11.4 provides a parallel description of states of the system as seen and recorded at different scales on different hierarchical levels. The values reported on the axes are not directly related to each other. It should be noted that the values taken by the various variables used as indicators of performance in the graph are not totally independent of each other within and across quadrants. For example, technical coefficients (throughputs per hectare of land and output/input ratios) and market variables (sale prices, structure of costs, and taxes and subsidies) define a direct link among many of the variables considered in the MCMSPS reading (e.g., economic return for farmers and environmental loading for the agroecosystem). This makes it possible to link many of these variables using equations of congruence across levels and tracking biophysical throughputs, economic flows, and profiles of human time allocation (Giampietro, 1997a, 1997c, Giampietro et al. 1994; Giampietro et al. 1998a; 1998b; Giampietro and Mayumi, 1997; Giampietro et al. 1997). In reference to farming system analysis, it is possible to frame a cross check in relation to land, human time, and money budgets (Giampietro and Pastore, 1999, Pastore et al., 1999).

This kind of mosaic effect (Prueitt, 1998) can be obtained by bridging encoding variables belonging to different scientific disciplines and various hierarchical levels in relation to the same set of variables. The flow of added value generated by a village has to be the result of the sum of the flow of added value generated by the lower level holons making up the village.

## 11.4.3 Dealing with the Problem of Moving across Hierarchical Levels

This peculiar aspect of the multi-criteria, multiple scale performance space deserves particular attention. When we represent the performance of the system at the household level, the quadrant describing the effects of farmers' choices on the environment (e.g., environmental loading) and on the socioeconomic context (e.g., the food surplus produced and its cost) refer to the specific and limited space-time scale at which the individual farm household is defined and described (e.g., a 200-ha farm in the U.S. over a period of 1 year, or a 1-ha Chinese farm over 1 year). To assess the effects of farmers' choices on a regional or national scale, one needs to aggregate the effects induced by the different choices made by individual farm households operating in a given village, region, or nation. Because the choices and the actions of individual farm households are not homogeneous, the problem of how to aggregate the effects of the behavior of individual farmers at higher hierarchical levels arise.

In order to move between levels, one can use a two step process (Giampietro and Pastore, 1999; Pastore et al, 1999). The first step is to define a set of farm types characterizing the typology of production in the area, resulting from the definition of accessible states for farmers. Such a set should cover a significant fraction of farmers behaviors (e.g., >90%). The second step is to define a curve of distribution of the population of individual households over the given set of existing types. After defining a set of farm types and specifying the characteristics of each type, we can obtain the aggregate behavior of a population of households over the next higher hierarchical level by considering how such a population is distributed over the set of types existing within the village. Clearly, a certain amount of information related to the characteristics of the village itself should also be added to the analysis. The land use within the village is not 100% determined by land use choices within farms; a certain fraction of the area occupied by the village is managed at the village level.

To obtain a set of characteristics that can be used to define a farm type, we can start by analyzing the constraints affecting farmers' options, as determined by internal links among the variables on the MCMSPS. For more details and numerical examples see Pastore et al., 1999. The steps of this process are as follows:

1. Choose the set of indicators of performance; these determine the skeleton of indicators for the MCMSPS.
2. Define a viability domain for each indicator (the range of values within which the farm can operate).

3. Define possible preferences of farmers in relation to different indicators; establish a preference relation among different areas within the viability domain.
4. Characterize the farm type in terms of selected combinations of farming techniques that saturate three endowments of the existing resources (accessible land, available labor time, and accessible financial capital), given the set of objectives defined in step 3.

Different strategies adopted by farmers (e.g., maximization of economic return or minimization of risk) can be studied as different profiles on the MCMSPS, as in the case of the preference of car buyers shown in Figure 11.3. The existence of internal constraints (e.g., a farm household cannot use more time, land, or capital than is available or accessible) implies that, given technical coefficients and the structure of prices, costs, and taxes, the possible choices for the farm household are limited. Studies of the nature of this limitation specifically address the peculiarity of research at the farm level as compared to research at the plot level (see Giampietro and Pastore, 1999; Pastore et al., 1999).

Each combination of techniques that satisfies the above mentioned conditions of (1) saturating as much as possible the existing budgets of land, labor time, and capital, and (2) operating within the selected set of indicators of performance, represents a viable technical option for farmers. Each combination is one possible state for the farm. Each farm type defined in this way implies a certain combination of trade-offs (a defined profile of values on the MCMSPS). This profile will determine a set of consequences not only for the farmers deciding to operate according to the characteristics of this type, but also for the environment and the national economy. Some farming types are more benign to the environment; others are more convenient for the society to which farmers belong; still others make possible a higher material standard of living for farmers in the short term.

## 11.5 STEPWISE APPLICATION OF THIS APPROACH

### 11.5.1 Selecting Indicators of Performance for Different Scales and Perspectives

The first step in using this approach is to select indicators of performance for each of the four quadrants: the household perspective, the socioeconomic (national) perspective, the environmental perspective, and the perspective relating to the system's ecological footprint.

A list of indicators that can be used to measure the performance of the system at household level is shown in Table 11.1. Assessments of the performance of a farming system at this level can consider various objectives, such as minimization of risk (e.g., safety from climatic, market and political disturbances), food security, maximization of income and net disposable cash, and maximization of the expression of potentialities for the members of the farm household (e.g., better education, better communication and information processing, and intensification of social and cultural events).

**Table 11.1 Indicators for Assessing Material Standard of Living at the Household Level**

| Indicator | Range of Possible Values |
|---|---|
| Average body mass | 34–60 kg |
| THT/C[a] | 10–45 |
| Dependency on market for food security | 0–100% |
| Endosomatic metabolic flow | 6.5–9.5 MJ/capita/day |
| Exosomatic metabolic flow | 35–900 MJ/capita/day |
| Net disposable cash | 50–50,000 US$/capita/yr |
| Average return of labor | 0.10–45 US$/hour |
| Expenditure for food | 5–75% of net disposal cash |
| Total food energy supply | 1500–4000 kcal/capita/day |
| Total protein supply | 30–130 g/capita/day |
| Animal protein/total protein ratio | 15–70% |

[a] THT/C = Total Human Time (total number of individuals belonging to the household × 8760 hours in one year)/Time (hours per year) allocated by the whole household for paid labor and subsistence chores.

Several indicators assessing agricultural performance from the perspective of the national or regional economy are listed in Table 11.2. At this level, several goals should be considered, such as selfsufficiency in food production, minimization of indirect costs of the food system, minimization of the direct economic cost of the food supply, and minimization of gradients in economic development between rural and urban areas.

Examples of indicators that can be used to monitor ecological impacts are presented in Table 11.3. The set of indicators should cover various distinct scales (e.g., world, region, watershed, village, farm, field, and soil). Again, an appropriate combination of these indicators depends on the scale and the type of information needed in the process of decision making.

Indicators that might be used in the fourth quadrant, which considers the degree of freedom from local biophysical constraints, are listed in Table 11.4. The goal is to compare the ecological footprint of the present agricultural system (the demands it places on natural resources and the ecosystem) with the natural resources and ecological services available in the physical boundary defined for the agroecosystem. A sustainable agroecosystem should be able to produce without generating irreversible deterioration in ecological systems. Indicators in this quadrant often represent the extent of linearization of matter and energy flows in the agroecosystem. The higher the rate of throughput on the farm, the higher the linearization of matter and energy flows in the agroecosystem; the greater the freedom from local natural constraints (technological inputs shortcut the ecological system of feedback controls), the greater the risk of generating negative consequences for the ecosystem (Giampietro, 1997a; 1997b).

## 11.5.2 Defining Feasibility Domains for Selected Indicators

Having chosen the variables on different axes distributed over different quadrants, one must define a range of feasible values for each indicator. In Figure 11.4, this

Table 11.2 Indicators for Assessing the Performance of Agricultural Systems According to Socioeconomic Context

| Indicator | Range of Possible Values |
| --- | --- |
| Average body mass | 34–60 kg |
| THT/C[a] | 10–45 |
| Dependency on importation for food security | 0–50 % |
| Exo-/Endosomatic energy ratio | 5–90 |
| Bio-economic pressure | 15–1600 Mj/hour |
| Exosomatic metabolic flow | 35–900 Mj/capita/day |
| Cereal surplus per hectare | –3000 to 4000 kg/ha arable land |
| Cereal surplus per hour | –1 to 85 kg/hr agric labor |
| Cost of agricultural surplus | –13 to 37 US$/hour labor |
| GNP/capita | 90–36,000 US$/capita/yr |
| Average return of labor | 0.10–45 US$/hour |
| Expenditure for food | 6–60 % of GDP |
| Total food energy supply | 1500–4000 kcal/capita/day |
| Total protein supply | 30–130 g/capita/day |
| Animal protein/total protein ratio | 15–70 % |
| Labor force in agriculture (%) | 4–70 % |
| Farmer income/national income average | 0.6–1.0 |
| GDP in agriculture/labor force in agriculture | 0.10–1.5 |
| Taxes from agriculture/subsidies to agriculture | (unpredictably variable) |
| Prevalence of malnutrition in children | 0.5–60 % |
| Infant mortality | 4–170 per thousand |
| Child mortality | 6–320 per thousand |
| Maternal mortality | 2–100 per thousand |
| Low birth weight | 4–40 % |
| Life expectancy | 39–79 years |
| Population/physician ratio | 210–73,000 |
| Population/hospital bed ratio | 65–65,000 |
| Pupil/teacher ratio | 6–90 |
| Illiteracy | 0.5–90 % |
| Radio ownership | 25–2,100 per thousand |
| Television ownership | 1–820 per thousand |
| Car ownership | 0.5–570 per thousand |

[a] THT/C = Total human time (number of individuals in the society × hours in one year)/time (hours per year) allocated by the whole society to labor in productive sectors of economy (food security, energy and mining, forestry and fishery, manufacturing).

range of values is indicated by the concentric shading. Light gray indicates a favorable value, dark gray an unfavorable value, and white an intermediate value. Within the feasibility domain we may add target values to the graph (the dots in Figure 11.4) that reflect the goals expressed by the stakeholders representing contrasting but legitimate perspectives.

The selection of indicators and their feasibility domains is a delicate and crucial step because according to the specific situations considered, there are always social groups not included in the participatory process (e.g., ethnic minorities, future generations, important stakeholders not recognized at the moment). If considered, they would introduce conflicting definitions of what is acceptable and that would require restarting the whole process (Giampietro, 1999).

**Table 11.3 Indicators for Assessing the Ecological Impact of Agriculture**

**Environmental loading**

kg of pesticides applied per hectare per year
kg of fertilizers applied per hectare per year
pollutants discharged into the environment

**Alterations of natural configurations of matter and energy flows**

indices of human alteration of gross primary productivity
thermodynamic indices of ecosystem stress
indices from theoretical ecology

**Bioindicators**

keystone species populations
plant associations
biodiversity assessment

**Landscape use patterns**

fractal dimension of agricultural landscape
hierarchical organization in space and time of matter and energy flows

**Table 11.4 Indicators for Assessing the Degree of Freedom of Agricultural Production from Local Biophysical Constraints**

| Indicator | Range of Possible Values |
|---|---|
| Output (endosomatic)/Input (exosomatic) energy ratio[a] | >50–0.1 |
| Indicators based on ecological footprint[b] | depends on the chosen indicator |
| Nutrient flows boosting ratio | 1–50 |
| Embodied land + Actual land/Actual land | depends on calculations |

[a] Measure of the dependency on fossil fuel energy.
[b] Natural capital required/natural capital available.

The existing linking of events across levels implies that dramatic changes occurring in the socioeconomic context within which the farming system operates will be reflected in the ranges of acceptable values on other levels. For example, a return of one dollar per hour of farm labor would be a remarkable achievement for a Chinese farmer, whereas such a return would throw farming in the European Union into a crisis.

## 11.5.3 Assessing the Current Situation of a Multidimensional State Space

In this step, the actual value of each indicator of performance in each of the four quadrants is recorded on a graph. This makes it possible to assess the values. Are they inside or outside their feasibility domain? How distant are they from their target? The multidimensional state space obtained at this point makes it possible to compare the current status of the system against the states defined as targets for policy implementation by stakeholders and against the feasibility domain based on the underlying biophysical links across hierarchical levels. Wide differences between actual values and expected values (either target values or values that would be required by congruence of matter and energy flows across levels) can be assumed to indicate stress in both natural and socioeconomic subsystems and indicate the need for intervention.

The two examples of a MCMS reading provided in Figure 11.4 refer to a standard characterization of farming in developing and developed countries. Figure 11.4a characterizes the situation of a subsistence farming system operating without external inputs. When population pressure is moderate, ecological indicators of stress are within the acceptable range, but the values of the set of indicators characterizing material standard of living are unacceptable according to any developed country based definition. The net disposable cash generated per hour of labor time, average body mass, and other social indicators of development are away from the viability domain at which rural households operate in developed countries. Figure 11.4b characterizes the situation of farmers operating in developed countries. In absolute terms, farmers in developed countries are better off than their subsistence farming counterparts. The multidimensional analysis reveals the trade-offs implied by this positive achievement on the socioeconomic side. Higher returns for humans in developed countries are paid for by the larger environmental impact of agriculture, by a heavy dependence on stock depletion (e.g., fossil energy), and often by import of ecological activity from distant ecosystems (e.g., imported animal feed and other agricultural commodities in Europe).

A comparison of the two profiles in Figures 11.4a and 11.4b (the distribution of actual results over the feasibility domains) shows the unbalanced negotiation among holons with contrasting perspectives when the farming system operates under different combinations of socioeconomic and ecological contexts. The ecological perspective tends to be the "loser" in intensive agriculture as soon as the demographic and socioeconomic pressures rise (Giampietro, 1997a,b). This explains why the cultural identity of traditional farmers undergoes heavy stress when fast socioeconomic development makes their traditional techniques no longer viable.

## 11.6 APPLICATION OF THIS APPROACH TO AGRICULTURAL INTENSIFICATION IN RURAL CHINA

We were able to identify three main farm types, each with minor variants, in an area of rural China:

- **Type 1:** Farmers who maximize net disposable cash (NDC) through cultivation of cash crops and off farm labor. On the negative side, this strategy means (1) taking risks because of the lack of self-sufficiency, (2) shouldering a heavy work load, and (3) creating heavy environmental stress.
- **Type 2:** Farmers who minimize their risk by growing mainly subsistence crops and maximize their leisure time (max THT/C) by avoiding off farm jobs. This strategy means remaining behind in the fast process of modernization of China,

as manifested by low net disposable cash and remaining on the unfavorable side
of a widening income gap between the average Chinese farmer and farmers using
these strategies.

- **Type 3:** Farmers who minimize risk by relying on subsistence crops and at the
same time attempt to maximize net disposable cash through off farm jobs and
cultivation of some cash crops. This strategy means heavy work loads (a low THT/C
ratio) and requires ample land and proximity of markets.

In our analysis, these strategies for using the available land, time, and capital
resources represent three different attractor solutions for the existing socioeconomic
and ecological context of the region and the cultural profiles of farmers. Clearly,
these farm types involve different trade offs in terms of performance in relation to
our four quadrants. Each produces a different MCMSPS profile.

The MCMSPS in Figure 11.6b shows that a Type 1 farm implies a higher income
for farmers along with the absence of a rice surplus to feed the urban population of
China. Actually these farmers are net consumers of rice, which is obviously bad for
the socioeconomic context. A Type 1 farm also generates a large and unfavorable
environmental load, which is obviously bad for the ecological context. These different implications of farmers' choices, one for the socioeconomic context and
another for the environmental context, are evident when comparing the MCMSPS
readings of Types 1 and 2. If Type 1 farms continue to spread througout rural China,
the country will no longer be able to feed its population without heavy reliance on
imports. Similar MCMSPS readings for other farm types are illustrated, discussed,
and assessed in Giampietro and Pastore (1999) and Pastore et al. (1999).

Each of these farming types which are defined at the household level can be
linked to a pattern of landscape use defined on the space scale of the farm. Considering the distribution of the population of farm households over the possible set of
farming types, we can calculate the characteristics of virtual villages made up of

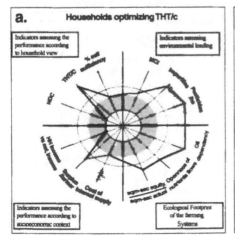
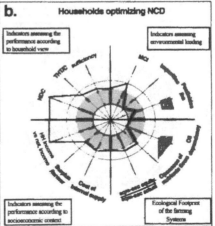

**Figure 11.6** MCMSPS readings for two different farm types: (a) type 2; (b) type 1.

different combinations of household types (both in terms of certain patterns of landscape use and aggregate effects on economic variables, such as the availability and the cost of rice).

To illustrate how our approach can be used to cross scales, we refer to Figure 11.7, which describe two villages simulated on the basis of the information obtained from the MCMSPS readings of farm types. The village described in Figure 11.7a is characterized by a majority of farmers who optimize NDC (80% of farmers belong to Type 1; 10% to Type 2; and 10% to Type 3). The village described in Figure 11.7b is characterized by a majority of farm households practicing traditional agriculture, hence minimizing risks and time allocated to work (10% of farm households belong to Type 1; 80% to Type 2, and 10% to Type 3). Note the different space-time scales of the MCMSPS readings. The scale of the village (Figure 11.7) is larger than that of the household (Figure 11.6). It covers a larger area, and therefore is slower in reacting to changes.

We could have generated simulated MCMSPS readings for households and for villages based on real data. This flexibility is one of the most powerful aspects of this approach. By doing both data collection and simulation at each level it is possible to validate the assumptions about farmers' behaviors adopted in a simulation. In China we found that the locations of villages (which affect access to markets and off farm job opportunities) were significant factors affecting the distribution of farmers over the three possible farm types. Farmers located far from urban centers are more likely to belong to Type 2. Similar hypotheses can be tested when considering population characteristics (age, sex, ethnic origin, level of education) as possible factors affecting the distribution of farmer households over the existing set of farm types. Young farmers are risk takers, more willing to keep in touch with the changes affecting the rest of Chinese society, and are therefore more likely to be Type 1.

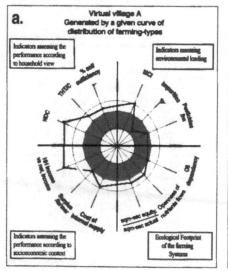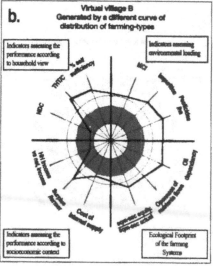

Figure 11.7 MCMSPS readings for two virtual villages.

From the MCMSPS readings presented in Figure 11.7, the first virtual village, where the majority of farmers do off farm work and engage in intensive production of cash crops, generates a much higher environmental loading than the second virtual village, it is more dependent on coal and oil for food production. From the national perspective, this village does not produce any surplus of rice; on the contrary, it erodes the rice surplus produced by nearby villages. What is detrimental to the environment and the food self-sufficiency of the country also has its positive side. A high level of net disposable cash for farmers and a lower risk of tension between rural and urban areas (a very sensitive topic for Chinese politicians) may result. The productive pattern adopted by Village 1 is benign to the villagers and to the people of the nearby town, who have access to a cheap supply of fresh vegetables and other food.

In contrast, the second virtual village (Figure 11.7b) has a surplus of rice (good for selfsufficiency of China) and generates a lower environmental impact than Village 1 (good for the environment). This environmentally benign village pays for these benefits with low net disposable cash from agriculture. People living in Village 2 risk being left behind by the dramatic socioeconomic transformation taking place in China. Expansion of the Village 2 type of farming will lock a large part of the Chinese rural population into a situation of poverty and lack of modernization.

We apply the same approach of scaling up to the interface between villages and the region or province. Given a spatial distribution of rural villages in a determined area and assuming several different distributions of rural villages over the set of possible village types, we can simulate changes in landscape use and effects on the economy of different changes in the distribution of village types in the area.

It should be noted that at each crossing of a new hierarchical level an external source of information about higher level characteristics has to be added to the process. The fraction of land used for common services in the village (e.g., a school) out of the direct control of farmers and the fraction of provincial land not under the direct control of villages cannot be determined by the analysis of the behavior of lower level types (characteristics of the types and the curve of distribution).

The larger the number of levels considered at the same time, the less reliable the mechanism generating simulations. In fact, when several levels are considered simultaneously (households, villages, province, country), it is easy to get into a situation in which changes in technology, farmers' attitudes, environmental settings, and governmental policies can feedback across levels, causing confusion. This may be due to the possible introduction of new farm types, the quick obsolescence of existing ones, or dramatic non-linear changes in the curves of distribution of lower level types over the set of accessible types.

## 11.7 CONCLUSIONS

In our view, the approach presented in this chapter provides a richer description of problems linked to sustainability in agriculture than do other existing methods of analysis. Because it is designed to overcome the shortcomings of other methods and deal more appropriately with the complex reality of systems of agricultural production, it can be an important tool for directing agriculture in a more sustainable direction.

To sum up the advantages of our approach, we believe it can provide a useful scientific basis for governance, decision making, and policy formation because it:

- Does not claim to provide the correct analysis of a system; rather, it generates several sets of view-dependent representations of the reality. The peculiarity of the approach is that it acknowledges such a dependency from the beginning;
- Can enrich policy making by including new alternative sets of view dependent representations and by enhancing negotiation among groups with different views and interests;
- Acknowledges that the goals related to the concept of sustainable development cannot all be achieved at the same time, and that it is impossible to adopt a single "silver bullet" technical solution;
- Recognizes that decision making implies finding compromise solutions through negotiation among legitimate but contrasting views;
- Enables the integrated use of information generated in different scientific fields (economics, sociology, agronomy, agroecology, theoretical ecology, etc.), as well as information that refers to nonequivalent descriptive domains (views of the same system on different space–time scales);
- Makes it easier to represent and discuss possible future scenarios;
- Possibly forces the consideration of the perspectives of stakeholders that normally are not included in the traditional analysis; and
- Makes possible the mandatory assessment of environmental costs on several space-time scales in the process of formulating policies affecting the sustainable use of natural resources.

## REFERENCES

Allen, T.F.H. and Starr, T.B., *Hierarchy*, University of Chicago Press, Chicago, 1982.

Bana e Costa, C.A., Ed., *Readings in Multiple Criteria Decision Aid*, Springer-Verlag, Berlin, 1990.

Brink, B.J.E., Hosper, S.H., and Colijn, F., A quantitative method for description and assessment of ecosystems: the AMOEBA approach, *Marine Pollution Bull.*, 23, 265–270, 1981.

Conway, G.R., The properties of agroecosystems, *Agric. Syst.*, 24, 95–117, 1987.

Giampietro, M., Sustainability and technological development in agriculture: a critical appraisal of genetic engineering, *BioScience*, 44(10), 677–689, 1994a.

Giampietro, M., Using hierarchy theory to explore the concept of sustainable development, *Futures*, 26(6), 616–625, 1994b.

Giampietro, M., Socioeconomic pressure, demographic pressure, environmental loading and technological changes in agriculture, *Agric. Ecosystems Environ.*, 65, 201–229, 1997a.

Giampietro, M., Socioeconomic constraints to farming with biodiversity. *Agric. Ecosystems Environ.*, 62, 145–167, 1997b.

Giampietro, M., The link between resources, technology and standard of living: a theoretical model, in Freese, L., Ed., *Adv. in Human Ecology*, JAI Press, Greenwich, CT, 6, 73–128, 1997c.

Giampietro M., Implications of complexity for an integrated assessment of sustainability trade-offs: participative definition of a multi-objective multiple-scale performance space. Paper for the European Commission Environment and Climate Programme Advanced Study Course, Universitat Autonoma Barcelona, September 20–October 1, 1999, in press.

Giampietro, M., Bukkens, S.G.F., and Pimentel, D., Models of energy analysis to assess the performance of food systems, *Agric. Syst.*, 45(1), 19–41, 1994.

Giampietro M. and Mayumi, K., A dynamic model of socioeconomic systems based on hierarchy theory and its application to sustainability, *Struct. Change Econ. Dynamics*, 8(4), 453–470, 1997.

Giampietro, M., Bukkens, S.G.F., and Pimentel, D., The link between resources, technology and standard of living: examples and applications, in Freese, L., Ed., *Adv. in Human Ecology*, JAI Press, Greenwich, CT, 6, 129–199, 1997.

Giampietro M., Mayumi, K., and Pastore, G., Socioeconomic systems as complex self-organizing adaptive holarchies: the dynamic exergy budget. Paper presented at the International Conference on Complex Systems, Nashua, NH, 21–26 September 1997. http://www.interjournal.org/#127

Giampietro, M. and Pastore, G., Multidimensional reading of the dynamics of rural intensification in China, the AMOEBA approach, *Crit. Rev. Plant Sci.*, 18(3), 299–330, 1999.

Giampietro, M., Bukkens, S.G.F., and Pimentel, D., General trends of technological changes in agriculture, *Crit. Rev. Plant Sci.*, 18(3), 261–282, 1999.

Glansdorf, P. and Prigogine, I., *Structure, Stability and Fluctuations*, John Wiley & Sons, Chichester, UK, 1971.

Hart, R.D. The effect of interlevel hierarchical system communication on agricultural system input-output relationships, Options Mediterraneennes Ciheam IAMZ-84-1, International Association for Ecology Series Study, 1984.

Ikerd, J.E. The need for a system approach to sustainable agriculture, *Agric. Ecosystems Environ.*, 46: 147–160, 1993.

Kampis, G. *Self-Modifying Systems in Biology and Cognitive Science: a new framework for Dynamics, Information and Complexity*, Pergamon Press, Oxford, 1991, 70.

Koestler, A., *The Ghost in the Machine*, Macmillan Compnay, New York, 1968.

Koestler, A., Beyond atomism and holism: The concept of the holon, in Koestler, A. and Smythies, J.R., Eds., *Beyond Reductionism*, Hutchinson, London, 192–232, 1969.

Li Ji, Giampietro, M., Pastore, G. Liewan, Cai., and Huaer, Luo., Factors affecting technical changes in rice-based farming systems in southern China: case study of Qianjiang municipality, *Crit. Rev. Plant Sci.*, 18(3), 283–298, 1999.

Lowrance, R., Hendrix, P., and Odum, E., A hierarchical approach to sustainable agriculture, *Amer. J. Alternative Agric.*, 1, 169–173, 1986.

Munda, G., Nijkamp, P., and Rietveld, P., Qualitative multicriteria evaluation for environmental management, *Ecological Econ.*, 10, 97–112, 1994.

Nicolis, G. and Prigogine, I., *Self-organization in Nonequilibrium Systems*, John Wiley & Sons, New York, 1977.

Nijkamp, P., Rietveld, P., and Voogd, H., *Multicriteria Evaluation in Physical Planning*, North-Holland, Amsterdam, 1990.

O'Neill, R.V., Perspectives in hierarchy and scale, in Roughgarden, J., May, R.M., and Levin, S., Eds., *Perspectives in Ecological Theory*, Princeton University Press, Princeton, 140–156, 1989.

Pastore, G., Giampietro, M., and Ji, Li., Conventional and land-time budget analysis of rural villages in Hubei province, China, *Crit. Rev. Plant Sci.*, 18(3), 331–358, 1999.

Prigogine, I., *From Being to Becoming*, W.H. Freeman, San Francisco, 1978.

Prigogine, I. and Stengers, I., *Order Out of Chaos*, Bantam Books, New York, 1981.

Prueitt, P.S., *Manhattan Project*, George Washington University BCN Group. http://www.bcn-group.org/area3/manhattan/manhattan.html

Rosen, R., *Anticipatory Systems: Philosophical, Mathematical and Methodological Foundations*, Pergamon Press, New York, 1985.

Rosen, R., *Life Itself: A Comprehensive Inquiry into the Nature, Origin and Fabrication of Life*, Columbia University Press, New York, 1991.

Salthe, S.N., *Evolving Hierarchical Systems: Their Structure and Representation*, Columbia University Press, New York, 1985.

Simon, H.A., The architecture of complexity, *Proc. Amer. Philos. Soc.*, 106, 467–482, 1962.

van den Bergh, J.C.J.M. and Nijkamp, P., Operationalizing sustainable development: dynamic ecological economic models, *Ecological Econ.*, 4, 11–23, 1991.

Wolf, S.A. and Allen, T.F.H., Recasting alternative agriculture as a management model: the value of adept scaling, *Ecological Econ.*, 12, 5–12, 1995.

Whyte, L.L., Wilson, A.G., and Wilson, D., Eds., *Hierarchical Struct.*, American Elsevier Publishing Company, Inc., New York, 1969.

# Index

9 780367 398118